# International Series in Operations Research & Management Science



Volume 322

More information about this series at https://link.springer.com/bookseries/6161

Igor Borisovich Shubinsky •
Alexei Mikhailovitch Zamyshlaev

# Technical Asset Management for Railway Transport

## Using the URRAN Approach

Igor Borisovich Shubinsky
JSC NIIAS
Moscow, Russia

Alexei Mikhailovitch Zamyshlaev
JSC NIIAS
Moscow, Russia

ISSN 0884-8289 ISSN 2214-7934 (electronic)
International Series in Operations Research & Management Science
ISBN 978-3-030-90028-1 ISBN 978-3-030-90029-8 (eBook)
https://doi.org/10.1007/978-3-030-90029-8

Translation from the Russian language edition: Management of Technical Assets of Railway Transport by Igor Borisovich Shubinsky, and Alexei Mikhailovitch Zamyshlaev, © Shubinsky, I.B., Zamyshlyaev, A.M. 2021. Published by Institute of Transport and Logistics Problems, LLC, printed by Nauka Publishing House. All Rights Reserved.



This Springer imprint is published by the registered company Springer Nature Switzerland AG.
The registered company address is: Gewerbestrasse 11, 6330 Cham, Switzerland

*Dr. Alexey Zamyshlyaev died suddenly from COVID-19. He was a talented person, an excellent specialist, and a promising scientist.*

# Preface

The book sets out the key provisions of the URRAN technical asset management system—a methodology for managing resources and risks by analyzing and ensuring the required levels of reliability and safety of railway transport facilities. The book also describes the architecture of the unified corporate platform of the URRAN information system (UCP URRAN) and its subsystems for infrastructure facilities and rolling stock complexes. The prospects of the development of the URRAN UCP are shown, especially in terms of the use of artificial intelligence to predict dangerous events in the operation of railway transport.

The book is intended primarily for specialists engaged in practical work on the technical maintenance of railway transport facilities. It is intended for students, postgraduates, and teachers at railway universities and can also be useful for specialists in other industries and transport, solving scientific and practical problems in the management of technical assets.

Moscow, the Russian Federation

Igor B. Shubinsky
Alexei M. Zamyshlyaev

# Acknowledgments

The authors would like to thank all JSC NIIAS experts who worked on the URRAN project as well as experts from design bureaus and departments of the JSC "RZD," railway representatives, and the management of the JSC "RZD" for useful discussions and taking part in the development of the URRAN system and its implementation in the Russian Railways.

# Contents

# Chapter 1
# Introduction

Well-balanced asset management is the basis for the success of any company. The company carefully treats its assets since its income and prosperity are determined by the condition of the company's facility with the use of which the profit is earned. Technical assets bring profit to the company only if they work reliably (smoothly) and efficiently. The efficiency of technical assets means the influence of their condition on the final result of the company's activity, i.e. on its profit. The efficiency of an asset is directly related to its productivity, functionality level, and especially safety. Thus, reliability, safety, productivity, and functionality are the main characteristics of a technical asset state. They have a decisive influence on the life cycle cost of an asset.

Tolerable levels of reliability and, mostly, safety are achieved on the basis of using a well-balanced system of technical maintenance of facility (technical asset of a company). This system should include advanced technologies of technical maintenance (TM) and facility repairs (R). Hence, the abbreviation for facility maintenance is TM and R. The well-balanced system of the facility technical maintenance is an integral part of the system for managing the condition of the technical assets of a company. The required levels of safety and reliability of the facility, as well as sufficient levels of their productivity and functionality, while ensuring an acceptable facility life cycle cost, shall be achieved through a balanced resource management based on a risk assessment. The solution to this problem is especially relevant for such a large backbone transport company in Russia as the Russian Railways that maintains an extremely large number of technical assets of infrastructure and rolling stock.

This monograph sets out the basics of creating a management system for the technical assets owned by the Russian Railways. This system is a risk-based system for integrated management of resources, reliability, and safety of Russian railway transport facilities (URRAN).

The relevance of the publication of the monograph is due to the fact that in the conditions of economic stagnation due to the COVID-19 coronavirus pandemic, the

 
I. B. Shubinsky, A. M. Zamyshlaev, *Technical Asset Management for Railway Transport*, International Series in Operations Research & Management Science 322,
https://doi.org/10.1007/978-3-030-90029-8_1

problem of rational management of the available resources of the railway company is particularly acute in order to ensure acceptable levels of dependability of facilities and technological processes, provided that the risks of traffic safety violation remain at an acceptable level. This problem has been solved and is being systematically implemented on Russian railways. Transport industry specialists should be directly involved in the operation of this system.

The text of the monograph is limited by the methodological provisions and the architecture of the information support for the URRAN system. The monograph includes the following sections:

- Introduction
- Problems of managing technical assets in railway domain
- Conceptual provisions for integrated risk-based management of reliability, safety, and resources
- Basic concepts and indicators of dependability and functional safety of railway transport facilities
- Standardization of the facilities of railway transport and normalization of dependability indicators
- Fundamentals of management of technical and industrial risks on railway transport
- Resource management of railway transport facilities
- Assessment of the activities of structural divisions of railway transport
- Unified corporate platform URRAN (UCP URRAN)
- Conclusion

# Chapter 2
# Problems of Managing Technical Assets in Railway Domain

## 2.1 General Description of the Problem

The railway transport of the Russian Federation is mainly represented by Joint Stock Company "Russian Railways" (JSC "RZD") that is the largest owner and operator of transport infrastructure facilities on the territory of the Russian Federation. Almost 335,000 people maintain the infrastructure complex of the JSC "RZD," including approximately 150,000 km of tracks, 30,000 bridges and viaducts, 159 tunnels, more than 5000 stations, and many other types of the infrastructure facilities. In addition, the JSC "RZD" is a major owner and operator of communication networks, an operator of telephone and radio communications, including digital communications (DMR, Tetra, GSM-R). The total length of communication lines of the JSC "RZD" is over 330,000 km; the length of fiber-optic communication lines is over 77,000 km. Over 500,000 units of signalling equipment, as well as more than 6 million units of various sensors, technical diagnostics and telemetry equipment are operated on the railway infrastructure. Transportation process is supplied uninterruptedly with electric power provided by 1402 traction substations. 11,000 freight locomotives (electric and diesel), 6000 shunting engines (diesel), 1,600,000 freight wagons of all types and owners, 24,000 long-distance passenger cars, and more than 15,000 commuter cars are operated on the railways of the JSC "RZD." The Russian Railways Holding is the largest backbone of the Russian economy, the most important element of its transport system, providing over 44% of freight turnover and over 30% of the passenger turnover of the entire transport system of the country, forming 1.7% of Russia's GDP, 1.5% of tax revenues to the country's budget system, and up to 4% of the total capital investment in Russia. Russian Railways is one of the TOP-5 largest companies in Russia, one of the top companies in the world (including the United States and China) in terms of the length of railway network and number of passengers carried and freight transported. The share of the cost of the infrastructure fixed assets of the JSC "RZD" is more than 60% of its total

 
I. B. Shubinsky, A. M. Zamyshlaev, *Technical Asset Management for Railway Transport*, International Series in Operations Research & Management Science 322,
https://doi.org/10.1007/978-3-030-90029-8_2

cost of fixed assets; the share of operational costs for infrastructure facilities is about 35% of the total costs. Optimization of infrastructure and rolling stock maintenance costs is one of the key objectives of the JSC "RZD."

The current situation in the world, associated with the introduction of wide-spread restrictions due to the COVID-19 coronavirus pandemic, has resulted in the stagnation of the global economy. The following circumstances forced the companies to abandon strategic development goals in favor of solving operational problems: stoppage in production, closure of state borders, a decrease in economic activity as well as population mobility up to zero, suspension of contracts for an indefinite period, and an increase in the risk of insolvency. As a result, it is impossible to develop long-term investment plans. Obviously, this situation is one of the most serious challenges to the existing world order and will require a radical revision of all existing principles of organizing production and running business, abandoning an unreasonable aggressive company growth policy related to borrowing a large amount of funds. Companies have to be moving on to conservative development strategy, relying only on available resources and borrowing a small amount of funds. In a period of the economy instability and an unpredictable market situation in the world, the issues of staying in business, keeping a team in place and its competencies as well as, if possible, holding a market share dominate.

All of the above problems have affected the transport market. The sector hit most directly was aviation sector. Aviation companies have to suspend business activities by canceling international flights and limiting the number of domestic flights. The railway transport, as a backbone of the state's economy, did not suspend its activities, but also faced a sharp decline in loading, a decrease in population mobility and, as a consequence, a decrease in revenue, which resulted in abandoning of strategic initiatives for the development of the infrastructure complex in favor of keeping the company's teams in place and ensuring current operational activities. So, according to the RBC (www.rbc.ru), JSC "RZD" recorded a decline in freight turnover by 7.3%, in March 2020 and in general, for the year it is expected to decrease to 5%. Passenger turnover is expected to decrease by 56% in 2020 and the amount of the investment for 2020 is expected to be reduced by 15%. The things are not going well in the European Union. According to the Global Railway Review (www.globalrailwayreview.com), rail traffic has declined by 30% since the introduction of quarantine measures. All participants of transport domain also have to hastily revise their investment plans in favor of solving operational problems.

At this particular time, it is extremely important to organize the work of railway companies in such a way as to ensure the maximum efficiency of all decisions taken in view of all aspects of production activities. Most nowadays railway companies are highly complex structures, where organizational hierarchies, processes, competencies, goals, performance indicators, and connections between people are tightly intertwined. That is why considering any problem, strategy, and initiative from only one point of view is not only unproductive, but also dangerous, since it only creates the illusion of solving the problem. The business model of a modern company in a simplified form is a continuous search of solutions to increase its profitability and reduce costs with the obligatory fulfillment of all the requirements of regulators (requirements for safety and reliability of train traffic, labor protection, environment safety, fire safety, etc.). Of course, it is impossible to unlimitedly reduce costs to achieve income growth. Therefore, the issues of finding an optimal strategy that would allow keeping a balance between costs, opportunities, risks, and asset productivity dominate for top management. In accordance with ISO 55000 standard an asset is item, thing, or entity that has potential or actual value to an organization. Thus, developing a modern effective organization management system is possible only on the basis of the principles of Asset Management that allow keeping a balance between its costs, performance, and risks assessment when choosing one of the development strategies. It should be noted that risk management should be carried out according to the principle that assessment of risks associated with large losses and a high probability of risk event occurrence is carried out first of all, and assessment of risks associated with smaller losses and a lower probability of risk event occurrence is carried out in descending order of risk event importance. This principle has especially proven its effectiveness when put into practice in large infrastructure companies with a geographically distributed network and requiring huge funds to maintain their infrastructure in serviceable condition to carry out their current operational activities.

## 2.2 Railway Company's Investments in the Technical Maintenance of Infrastructure

The railways around the world have to invest a prodigious amount of money resources in the development and maintenance of their infrastructure. In European countries, the cost of the maintenance of the track superstructure amounts to about 50% of the cost of its life cycle. The volume of the world market for maintenance and repair of rolling stock is currently 6% higher than the value of the market for new equipment. The current volume of the world market for after-sales services for rolling stock is estimated at around 54 billion euros. According to experts, the market size will grow on average by 3.2% annually until 2022.

Following the lifting of quarantine restrictions, European Railways have to continue its operation maintaining a high level of business efficiency while carrying out a program of economic austerity measures. Even more difficult is the fact that in some cases the railways are owned by different companies. British experts (Gill Plimmer and Jonathan Ford,—the Financial Times article "A fragmented Network," 01/30/2018) analyzed the results of the privatization of the national railways of the UK, which took place since 1992. The authors of the article concluded that the model of privatization of the railway industry with the fragmentation of a single system in the past it into three components of infrastructure, rolling stock and train operators resulted in increasing the cost of infrastructure maintenance. The cost of running the UK's railways is about 40% higher than it is in the rest of Europe at different times since each player in the railway transport services market is focused on their business issues, and not on interaction with other market players and thus improving the operation of railways in general. The subsidies that were received by the infrastructure managers insulated private operators from the extra costs incurred by the national rail network. While it cost £4.1bn to provide maintenance and renewals work on the infrastructure in 2016–17, the train operators paid £1.5bn to access the nation's tracks. The train companies having operating margins of 3% and limited term of the contract (franchising system), and contributing very little in the business paid £634m as dividends out of nearly £688m operating profits between 2012 and 2016. Thus, in the expert and public community, the opinion was formed that the UK's attempt to create a unique model of competitive relations on the railway network did not succeed in many aspects. While privatization has resulted in increasing the costs, it has failed to achieve the goals of upgrading infrastructure and stabilizing its financial position. Today, Network Rail (NR), the owner and operator of rail infrastructure in the UK, which was created in 2002 to replace the liquidated state-owned Railtrack, plans to spend \$ 45.17 billion in 2019–2023 of which \$ 9.9 billion is for infrastructure maintenance and \$ 21.4 billion is for infrastructure upgrades.

The experience of France is also remarkable. After 17 years of separation of infrastructure from transportation service, SNCF and RFF were reintegrated in 2015. The model of horizontal separation, when a single management focuses on issues of both infrastructure and transportation activities mostly allows avoiding the above problems. On average, SNCF spends 2.6 billion euros annually on infrastructure maintenance.

German Railways (DB) is also increasing the costs on infrastructure maintenance. So, until 2014, the budget for the maintenance amounted to 1.4 billion euros per year, since 2015 it was increased to 1.6 billion euros per year, in subsequent years it was increased to 1.9 billion euros per year. In 2020, the budget amounted 2.4 billion euros per year. At the same time, there is still underfunding of some of the necessary infrastructure works, in particular, this applies to a number of large bridges built in the second half of the nineteenth century.

The infrastructure operator ÖBB Infrastructure (Austria) invested 1.8 billion euros in its assets in 2017, and, according to the company's plans, this figure will increase to 2.6 billion euros in 2021.

Italian Railways (RFI group of companies) intends to invest about 5.5 billion euros in the maintenance and diagnostics of infrastructure in order to improve the punctuality of train movements during 2019–2023.

The volume of investments spent by railway companies for the technical maintenance of infrastructure and rolling stock is quite considerable, and is one of the most important categories of costs in the total cost of the life cycle of a technical asset. Thus, developing an effective strategy for the technical maintenance of infrastructure and rolling stock is a key task for any railway company.

## 2.3 Projects of the International Union of Railways (UIC)

Determining the necessary costs for the maintenance and upgrading of the railway network, taking into account the evolving requirements of both users of this mode of transport and other stakeholders, is a challenging task for governments and regulators in many countries around the world. Given the importance of this issue, the International Union of Railways began collecting and analyzing relevant data as early as in 1996 within the framework of the Lasting Infrastructure Cost Benchmarking (LISB) project. The infrastructure managers from 14 European countries have been participating in the project for over 20 years. The aim of the project is to determine the costs needed for railway companies to maintain and upgrade the existing infrastructure, as well as to determine the factors that influence these costs. It was found that in some cases the higher load on the rail network increased maintenance costs by about 5% and upgrades costs by more than 16%. The LISB project became a starting point for the accumulation of statistical data on the technical maintenance of infrastructure, the accumulation and in-depth analysis of the best practices of railways, and the exchange of experience between the project participants.

In 2007, the UIC established the Asset Management Working Group (AMWG), which included representatives from 10 European railway companies. Russian experts joined this group in 2016.

As part of the Group's activities the document "Guidelines for the Application of Asset Management in Railway Infrastructure Organizations" was developed in 2010 and is regularly updated in accordance with results of the annual Group activity. In 2015, a short list of Key Cost Drivers in Railway Asset Management was published. Based on the statistical data of the project participants, the influence of various cost factors (the so-called value drivers) on the cost of maintenance, repair and upgrade of infrastructure was determined. It was noted that there are quantitative and qualitative value drivers. Quality drivers could not be quantified, so its descriptions were presented in the document. In 2016, UIC issued practical guidance on the application of the ISO 55001 standard in the field of railway asset management. The Russian participation in particular contributed to this international document with the experience accumulated by Russian Railways during the development and application of the Integrated system of dependability, risks, and resources at all lifecycle stages, called URRAN for short. Special attention in this document is paid to the issues of risk management. This document, like no other, allows you to ensure the implementation of the basic principles of asset management: "reducing costs by doing the right work in the right place at the right time with interventions coordinated to achieve the optimum balance between maintaining, renewal and enhancement across the asset base."

In 2020, the Group initiated a new project "Asset Management Whole System Decision Making" (WiSDoM). The aim of the project is to develop a concept and related methods and tools for a Unified Decision Making System (hereinafter—the "System"), designed to provide asset management in railway domain. According to the concept of the System it is assumed the integration of the "system of systems" approach used in other industries with existing approaches to asset management is close to the URRAN methodology in many ways. The project implementation plan provides for the development of a concept for the application of the System for the entire infrastructure, unified processes and decision-making criteria, the definition of methods and tools to ensure decision-making, practical testing, validation of the applied methods, and confirmation of the effectiveness of the System. The project implementation period is from January 2021 to December 2023. In addition to the permanent members of the Group, representatives of multidisciplinary international consulting organizations, research institutes and transport companies from Austria, the USA, the UK, Ireland, France, the Netherlands, etc., take part in regular meetings devoted to asset management.

## 2.4 Application of Digital Technologies in Asset Management Systems of European Railway Companies

As noted above, one of the main tasks of any railway company is to develop an effective strategy for managing the technical condition of infrastructure and rolling stock within an acceptable level of costs. This is especially vital in dynamically changing circumstances of the post-COVID world and low oil prices when the established economic relations between countries are being replaced by others due to new societal demands in essential commodities, medicine and equipment and due to market conditions. Nowadays, there appear new logistic supply chains and transport routes, respectively, that have not been used before. In this context we may expect a demand increase for transportation in one directions and a downright cut in others. Such a disproportion can bring a long line of serious problems for railway companies; in particular, they have to invest much in renovation and improvement of their infrastructure, to maintain it in good operating state while supporting a growing volume of transportation. So, railway companies are interested in developing strategies for efficient asset technical condition management that allow increasing transportation volumes with a high level of safety and dependability based on Big Data of diagnostics systems. As noted in ISO 55001: 2014 "objective information and knowledge about the condition of assets, their performance, risks and cost, and their relationship is vital for successful asset management." In order to efficiently manage railway infrastructure and rolling stock, one shall know their current technical state and be able to predict their state in future.

At present there are a lot of talks about the demand for implementation of disruptive (including digital) technologies on railways. Basically, the need for breakthrough technologies in railway transport was constantly present. It should be noted that railway transport itself was the product of innovative ideas at one time. In this case, the question is what should be considered innovation and how specific innovative products should be incorporated into the day-to-day operations of railways. Besides obvious benefits, digitalization of the asset management system can bring some serious problems for railway companies. It is possible in those cases when a grade of automation has been wrongly chosen, or wrong parameters have been used for generation of a database of technical assets. Problems arise especially in cases when a company has no objective information about a current real state of technical assets, and available information is just some irrelevant reports produced by quasi-automated systems that are operated in line with the processes described by hired business consultants.

It is also very important to have a clear idea about the size of planned financial expenditures. Advanced technologies of data collection, transfer, and analysis in the nearest future can radically change the maintenance rules adopted today on railways and in principle based on a normative approach. Monitoring systems for technical facilities of infrastructure and rolling stock are becoming more available, and they generate huge amounts of data that has been called Big Data, Internet of Things (IoT), Internet of Services (IoS), and machine self-learning—all these concepts are actively taking their places in short-term digital strategies of railway development. Today the hot subject of researches is becoming the search for a balance between a rate, process costs, data transfer speed, and power consumption of such multisensory systems. As a small example, there is DIANA system developed by infraView for the purpose of diagnostics and analysis of DB infrastructure components. In 2016 over 6500 switches were connected to it, while in 2019 they already amounted to 25,000, and in 2020 the system controls 30,000 of switches. Even today such smart systems begin to generate very huge amounts of data. Their transfer, storage, processing, and analysis will become important issues in the years to come. Therefore, there arises a question whether these Big Data are consolidated, analyzed, and used by railway companies in the right way. As when using prediction, it is difficult to assess whether various risk levels are acceptable and to take measures that would have seemed too risky before. For instance, one may reduce redundancy and cut the costs of technical maintenance if a decision-making person believes it acceptable when a wheel pair achieves a limit state prior to a scheduled repair, though previously he would immediately have sent it for repair. However, while balancing at the level of risks, one should understand that decision makers will also have to balance at the level of responsibility for risk-related decisions made. Therefore, railway companies will have to take a lot of efforts so that the huge investments that they are currently putting into digitalization should not be lost, since the implementation will face opposition from the established system of administration and distribution of responsibility among management levels within a company.

In this context, some positive experience accumulated by the Federal railways of Austria OBB can be useful for other railway companies. OBB strives to take a lead in implementing ISO 55000 on railway transport and successfully verifies this intention in practice, by getting the ISO 55000 asset management certificate the first among all European railway companies in February 2019 for shifting to track maintenance based on the following principles:

- Periodic full renovation of track
- Application of a current maintenance program combining preventive measures and those based on information about a real track state received by regular inspections
- Cancellation of permanent speed restrictions on mainlines
- Rehabilitation of roadbed when necessary on mainlines with dense traffic and at locations of switches.

The company experts have developed a unified network plan of asset management detailed for each technical department.

The OBB administration pays a lot of attention to the issues of IT support for the asset management system as a major tool for implementation of this methodology. Thus, from 2013 till 2018, the OBB management allocated over 12 mln Euro for the development of an IT platform. As a result of the work, Austrian engineers integrated 360 various databases into one database in order to construct an infrastructure asset management system.

Italian railways RFI became a second European railway company who was certified for their asset management system. This has been the first and the only certification in Italy accredited by Italcertifer through Accredia for physical assets management. The cost of the certification for RFI was over 40000 Euro. As the project manager Donatella Fochesato said, during the certification procedure 104 processes and 387 sub-processes of the company were described.

The certification is part of a wider strategy aimed at further improvement of network management and creating the necessary product for the company as well as for concerned parties.

Irish Rail has invested over 20 mln Euro in developing an IT system of asset management.

Belgian railways Infrabel has spent over 10 mln Euro for the development of an IT system for asset management.

The generalized architecture of the European asset management IT platform is presented in Fig. 2.1. It provides rather a diverse spectrum of IT solutions, which project managers in charge of introduction of an asset management in companies admit to be a big problem as making any changes to the platform's software is very time-consuming and requires a lot of financial resources and interactions with an IT department and a software supplier.

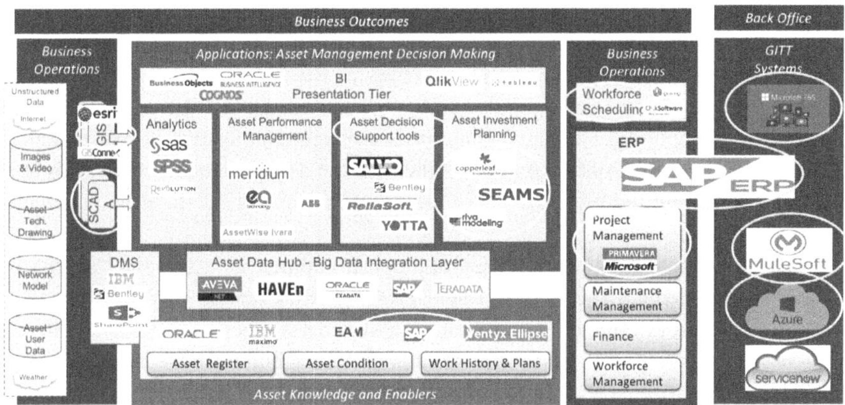


Fig. 2.1 Generalized architecture of the European asset management IT platform (authored by Jude Carey, Irish Rail)

## 2.5 Certification of the Asset Management System for the Subdivisions of Russian Railways for Compliance with ISO 55001: 2014

The object of certification for compliance with the requirements ISO 55001: 2014 is the asset management system—the constant and coordinated activity of the organization aimed at optimal management of assets, their productivity, risks and costs at all stages of the life cycle to fulfill strategic plans of the organization.

An asset management system includes a set of interrelated elements—organizational structure, functions and responsibilities of personnel, action plans, documentation, information systems, methods, processes, procedures, and resources. These elements are already available in the subdivisions of the JSC "RZD," since their activities are sufficiently formalized and documented. An asset management system is primarily developed and implemented in support of the organization's strategic plan and its strategic goals. The organization's strategic plan is the starting point for the development of asset management system documents—asset management policies, strategies, goals, and plans. The JSC "RZD" has a number of strategies in specific areas. Some of them are directly related to asset management issues, for example:

- Real estate management strategy of the Russian Railways Holding
- Strategy for managing the intellectual property of the Russian Railways Holding
- Strategies for managing the branding assets of the Russian Railways Holding.

One of the strategic goals of the Company in accordance with the Development Strategy of the Russian Railways Holding for the period up to 2030 is: "To ensure the systematic renewal of assets using innovative technologies and solutions based on effective management of the life cycle cost, availability and reliability of fixed assets" (this chapter). The Strategy also determines that "the Holding considers personnel as its most important asset" (item 5.5). Thus, the strategic goals related to the Company's asset management activities at Russian Railways have already been established.

One of the important elements of an asset management system is risk management, monitoring, and continual improvement of the asset management system (for example, increasing the productivity of assets or optimizing the asset management plan (s)).

Strategic risk management of the Company is implemented within the framework of the risk management and internal control system. Risk management for technical assets is carried out in branches and their structural divisions in accordance with the methodology of Resource, Risk, and Reliability Management for railway transport facilities at life cycle stages (URRAN), within which both methodological documentation and software for automating the management process have been developed. It should be noted that since 2010, JSC Russian Railways has started to develop and introduce a methodology, a set of standards, and guidelines for the management of technical assets of railway transport based on the URRAN methodology. To that end, Russian railways initiated the process of harmonization of the Russian infrastructure management regulatory framework with the RAMS standards widely used by the EU and US railway companies [1, 2, etc.]. RAMS is a methodology for ensuring Reliability, Availability, Maintainability, and Safety on railway transport. This is a corporate effort of the European Union formalized by the standards EN50126/IEC62278, EN 50128/IEC62279, EN50129/IEC62425Ed, and EN50159/IEC62280 (first and second part of the standard). RAMS targeting at manufacturers of technical equipment did not satisfy the goals of JSC RZD which are focused around operational activity. In addition, the RAMS methodology is developed at the level of individual objects and does not affect the processes of reliable and secure asset management of enterprises in accordance with current asset management standards GOST R 55.0.01/ISO 55000: 2014 [3, 4]. Currently, these standards are conceptual. Therefore, for their practical development and taking into account the results of harmonization with the RAMS standards, new methodological directions have been developed for the effective maintenance management of the Company. As a result, a technical asset management system was created based on the URRAN methodology. This system includes a set of international and national standards, organization standards, methodologies, methodological recommendations and, which is very important, information support for the railway network.

Completing the certification of the asset management system confirms that the system is introduced, complies with the basic level established by the requirements of ISO 55001, and allows the efficient management of the company's assets. In view of this, JSC Russian Railways can get the following advantages:

- Improving the image in the transportation market
- Increasing the systemic stability of the business
- Attractiveness for investments and obtaining borrowed funds
- Increasing the level of trust of stakeholders, including government authorities

- Guarantees of customer satisfaction and stability for partners
- Demonstration of social responsibility and attractiveness for potential employees (if personnel are included in the asset management system as one of the types of assets)
- Growth of income, production and commercial efficiency of the holding.

ISO 55000 certification is also used as a competitive advantage in railway tenders.

Summarizing all of the above, the following conclusions can be drawn:

- Russian Railways has an asset management system that, to a certain extent, meets the requirements of the ISO 55001 standard
- To achieve full compliance with the standard, it is necessary to formalize and adjust the existing system to meet the clear requirements of the ISO 55001 standard, such as defining asset portfolios, preparing asset management system documents (policies, plans), selecting processes and procedures used in asset portfolio management, training experts, and conducting training in the field of asset management
- It is reasonable to carry out certification both for individual branches and for the Russian Railways as a whole.

ISO 55001: 2014 certification is voluntary. The following voluntary certification systems are recognized in this area in the Russian Federation today:

- DQS—international audit and certification holding
- Certification Association "Russian Register"
- NAUCERT established on the basis of LLC Interregional Scientific Center for Comparative Research and Conformity Assessment.

Certification for compliance specifically with the ISO 55001 standard with the involvement of an experienced foreign accredited certification body is more reasonable for JSC "Russian Railways." In this case, the certificate will be recognized not only in the Russian Federation, but also in Europe and the world, which will facilitate the participation of Russian Railways in international tenders.

It should be noted that according to the International Technical Committee for Standardization ISO/TC 251 "Asset Management",[1] more than 280 organizations in the world are currently certified according to ISO 55001: 2014, among which more than 12% of the companies are from transport (among them are railway organizations—SMRT Rail (Singapore), Downer EDI Rail (Australia), Sydney, London, and Atlanta Metro). The company "Russian Railways" could take its rightful place among the above companies.

[1] According to https://committee.iso.org/sites/tc251/social-links/resources/known-certified-organizations.html

## References

1. IEC 62278:2002 Railway applications – Specification and demonstration of reliability, availability, maintainability and safety (RAMS)
2. Patra, A.: RAMS and LCC in Rail Track Maintenance. Division of Operation and Maintenance Engineering, Luleå University of Technology (2009)
3. Iorsh, V.I.: Upravlenie aktivami v sootvetstvii so standartami ISO (Asset management in accordance with ISO standards). Upravlenie Kachestvom Journal. 5 (2015)
4. The Asset Management Landscape, 2nd edn. Available at: https://gfmam.org/sites/default/files/2019-05/GFMAMLandscape_SecondEdition_English.pdf

# Chapter 3
# Conceptual Provisions for Integrated Risk-Based Management of Reliability, Safety, and Resources

## 3.1 Management of a Railway Transport Facility—Management of a Physical Asset of an Organization

In accordance with ISO 55000 series an asset is an identifiable item, thing, or object that has potential or actual value for the organization. Value can be defined in many ways by different enterprises and their stakeholders and can be tangible or intangible, financial or non-financial and include risks and liabilities. The value can be positive or negative at various stages of the asset's life. Physical assets usually include equipment, inventory, and real estate owned by an organization. Physical assets are the opposite of non-physical intangible assets such as rights to use intangible objects, brands, digital assets, intellectual property rights, licenses, reputation, and business relationships. A group of assets that make up a system of assets can also be considered an asset (ISO 55000 series).

Asset management contributes to the creation of value by balancing financial costs, environmental and social costs, risks, quality of service, and productivity related to assets. Asset management does not focus directly on the asset itself, but on the value that the asset can provide to the organization. Value (which may be tangible or intangible, financial or non-financial) is determined by the organization and its stakeholders in accordance with the goals of the organization. Asset management requires accurate information about assets, but an asset management system is more than an information management system. The assets can also provide more than one function and be used by more than one structural division of the organization. In accordance with the ISO 55000 series asset management is supposed to find a balance between costs, opportunities, risks, and the required productivity of assets. Objective information and knowledge about the condition of assets, its performance, risk and value and their relations is vital for successful asset management.

The asset management system implements a iterative management cycle (the so-called Deming cycle), including planning, doing, analysis of results (checking),



I. B. Shubinsky, A. M. Zamyshlaev, *Technical Asset Management for Railway Transport*, International Series in Operations Research & Management Science 322,
https://doi.org/10.1007/978-3-030-90029-8_3

and improvement of the organizing the activities (acting). This management system is supposed to have a clear separation of powers with a leading role of management, uses input data (organization context), and includes support tools (resources, organizational structure, information systems, data and information about assets, competencies). Decisions made at the beginning of the asset's life may not be optimal at the end of the asset's life. Asset management involves finding a balance between costs, prospects, and risks, on the one hand, and ensuring the required productivity of assets, on the other hand, to achieve the goals of the organization.

The following become mandatory in the asset management system:

- Information (data on assets)
- Technical condition assessment
- Risk assessment
- RCM-process, where RCM is reliability-centered maintenance
- Life cycle cost analysis
- Analysis of efficiency indicators, including overall equipment efficiency (OEE), etc.

The use of an integrated approach to enterprise management allows the enterprise asset management system to be formed using elements of the existing systems: quality management, environmental management, occupational safety and health protection, risk management, etc. Creation of the system on the basis of existing systems allows to reduce the efforts and costs associated with the development and maintenance of the asset management system in a satisfactory state of operation. It also improves the integration of various functions and the coordination of the activities of the structural divisions of the enterprise.

An enterprise has to have the appropriate competencies for effective asset management [1]. Specifically, it has to have personnel with appropriate responsibilities and competencies in the required areas. In accordance with the GFMAM international consensus [2], a number of disciplines are identified that form the structure of the necessary competencies in asset management.

These disciplines are combined into six groups:

- Development of management strategy and planning of asset management
- Making decisions on asset management (capital investments, operation and maintenance, etc.)
- Activities at the stages of asset life cycles: creation and acquisition of assets, systems engineering and management of asset configuration, fulfillment of maintenance and repair
- Risk and evaluation in the process of asset management: criticality, risk management, emergency preparedness, management of changes, monitoring of results achieved, interaction, reliability, and analysis of failure causes, emergency response, modernization, etc.
- Information systems for asset management; data and knowledge about assets, etc.
- Organizational structure and people: procurement and supply management, leadership, organizational structure and culture, competence and behavior.

A technical asset management system is a set of interrelated and interacting elements of an organization for the development of a technical asset management policy, for the asset management goals and the processes that are necessary to achieve these goals.

The requirements for the asset management system set out in ISO 55.0.01 are structured in such a way that comply with the basic principles of asset management: the context of the organization; leadership; planning; support tools; functioning; assessment of results; improvement.

## 3.2 Basic Principles of Integrated Management of the Technical Maintenance of Railway Transport Facilities

Within the framework of the concept [3], a phased approach to the introduction of integrated management of reliability, risks, life cycle cost in railway transport is determined. The main goal of the concept is to contribute to the achievement of a common understanding and development of common approach to the management of reliability and safety indicators of railway transport facilities in terms of assessing risks and life cycle costs.

The issue of the effectiveness of financing maintenance and repairs raises difficult questions for the chief executive officer of any company:

- How do the total costs of the organization depend on the amount of funds invested in the maintenance (technical maintenance and repair) of the equipment? What should be done? Should one reduce or, conversely, increase maintenance costs in order to reduce the overall costs of a company, including repair costs and equipment downtime losses?
- How to make an effective decision on how much and in what areas it is possible to reduce costs with an acceptable decrease in the level of equipment reliability?
- Whether a decision based on incomplete or insufficient data result in losses due to an increase in a number of downtime events and the scope of expensive repairs?

The solution to these issues is achieved by:

- Ensuring a complex and coordinated activities of the organization
- Management of physical assets and their modes of operation, risks and costs during the life cycle to fulfill the strategic plans of the organization
- Reducing equipment downtime, reducing the cost of maintenance, repairs, and procurement costs.

The advanced strategy of maintenance and repair was formed as a result of the change of the following four generations of its development [1]:

- The 1930s—Emergency-Recovery Works (ERW). Principle: "We fix it when it breaks." The first generation of a maintenance and repair strategy. Advantage:

relatively short service personnel training time. Disadvantage: high failure rate of the device.

- The 1950s—Scheduled Preventive Repair (SPR). The principle of planned terms of service and repair. Second generation of maintenance and repair strategy. Advantage: Reduction of equipment downtime and reduction of failure rate. Disadvantages: excessive frequency of maintenance, increased costs, and increased cost-of-ownership of assets.
- The 1980s—Condition and Reliability-based Planning (CRP) (Reliability-Centered Maintenance—RCM). The principle of maintenance and repair according to the state of the system. Third-generation maintenance and repair strategy. Advantages: Decision-making is based on the state of assets. Reduction of failure rate. Reduction of system maintenance frequency. Disadvantages: There is a large share of the experts evaluation component in assessing the condition of the equipment. Adoption of unbalanced decisions in terms of "income–risks–costs" for the long term. The second and third generations of maintenance and repair strategies are standardized by GOST R 53480 (maintenance philosophy) and GOST 15.601 (maintenance and repair strategy—scheduled maintenance, maintenance with periodic monitoring of parameters, maintenance with continuous monitoring of parameters, scheduled repair, technical condition-based maintenance).
- 2000–2020—Asset Ownership-based Planning (AOP)-Results Based Management (RBM). A reliability management principle taking into account the cost-of-ownership. The fourth generation of the maintenance and repair strategy. Advantages: Decision-making is based on the state of assets. Improved accuracy of forecasts of the equipment condition and failures. Management of risks, revenues, and costs at all stages of the asset life cycle. Effective investment solutions for long term. Investments in the development of technical personnel and information systems. Disadvantage: additional investments in the technical personnel training and development of information systems.

Maintenance and repair strategies are shown in Fig. 3.1, where EAM stands for Enterprise Asset Management. Condition and reliability-based Planning (Reliability-Centered Maintenance—RCM) (CRP-RCM) is a process of developing and making decisions aimed at identifying suitable and effective requirements for the system and preventive maintenance operations. Such requirements should take into account the consequences of detected failures in terms of their impact on the safety, technical efficiency, economy of operation of the product, and the mechanisms of its degradation that cause these failures.

This strategy allows reducing costs by eliminating redundant preventive maintenance or repair of equipment whose failure is of low criticality; focusing resources on critical components of the production system; improving the reliability of the production system, reducing the risk of damage to production, safety and the environment; shifting the focus of attention from the question "how to avoid failure?" to the question "how to avoid the consequences of failures?" Reliability-centered maintenance involves defining a set of necessary measures to ensure that

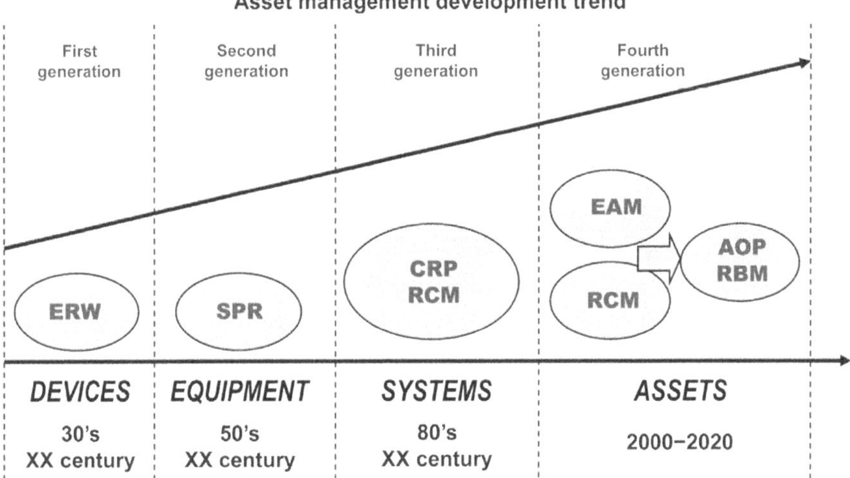


**Fig. 3.1** Maintenance and repair strategies

any facility continues to perform the functions required by the owner in the current operating situation (operating conditions).

In accordance with RCM each facility is formally described by answering the following key questions: What are the functions of the facility (what does the user need)? In what way can a facility fail? What can lead to failure (cause of failure)? What happens when a failure occurs (failure result)? How vital is failure (consequences of failure)? Is there anything you can do to predict or prevent failure? What if the failure cannot be predicted or prevented?

Depending on the answers and according to the typical decision-making scheme, measures are determined to prevent failures, in other words, the proposed strategy for the maintenance and repair of equipment is selected.

Overall, reliability-centered maintenance is characterized by:

- Focus on increasing the level of personnel and environment safety
- Increasing the economic efficiency of the use of capital assets
- Increasing the service life and productivity of the equipment
- Shifting the focus from equipment repair to reducing the number and mitigating the consequences of failures
- A high level of information support of decision-making processes.

Reliability-centered maintenance is an increasingly popular maintenance and repair strategy that is gaining more and more supporters around the world. The CRP-RCM strategy seamlessly combines the developments and achievements of the previous stages, while ensuring the required level of equipment operability and minimizing the costs of its technical maintenance.

As a result of the evolution of the reliability-centered maintenance strategy, a risk-based maintenance strategy has emerged.

A key feature of a risk-based maintenance strategy is the introduction of another parameter—the probability of equipment failure. Simultaneous consideration of the equipment criticality rating (in accordance with Reliability-Centered Maintenance strategy) and the probability (frequency) of its failure, which can be determined on the basis of the available failures statistics or by an expert method, makes it possible to identify the level of risk related to the operation of the equipment. Availability of information on the risk level makes it possible to balance the terms and volumes of maintenance and repairs of equipment, setting the priority of their performance, which in general leads to an increase in the reliability and safety of the production complex, as well as to a decrease in associated costs.

A risk-based strategy includes:

- Estimation and forecast of the probability of equipment failures
- Obtaining a quantitative assessment of the consequences of failure and calculating risks
- Calculation of the total risk, including the accumulated risk over a period of time
- Determination of the ratio of risks to the costs of reducing them and making a choice between eliminating and accepting risks
- At all stages of the life cycle, selection of priorities in maintenance includes planning of activities with ranking, comparing the necessary resources for activities with resource constraints, focusing the available resources on the top-ranked activities that will ensure the greatest reduction of the overall risk
- Minimization of costs for technical maintenance while maintaining an acceptable level of risks and accomplishing the strategic goals of the company.

The strategy of scientific and technical development of JSC Russian Railways for the period up to 2030 (White Book of JSC Russian Railways) defines the guidelines for the innovative development of the company. One of these guidelines is the creation of conditions for a stable, safe, and effective functioning of railway transport as backbone of the country's transport system in order to achieve the main geopolitical and geo-economic goals of Russia.

The state documents on the strategic development of railway transport (including the Federal Law "On Strategic Planning in the Russian Federation" and the Transport Strategy of the Russian Federation for the period up to 2030) determine the technical and production parameters for the development of railway transport upon transition to an innovative and socially oriented type of economic development. In accordance with these documents the country's transport complex was set an ambitious task of creating a system for managing the technical maintenance of railway transport facilities based on an advanced risk-based strategy.

The "White Book" of JSC Russian Railways emphasizes that "Safety of railway transport remains the main priority of the strategic scientific and technological development of the Russian Railways Holding. However, in the context of restrictions on deliveries of high-tech products to Russia and potential shortage of spare parts and materials supplied by foreign companies, repair and modernization of technical facilities and systems (that were previously assigned according to the service life or handled tonnage, and often without taking into account the previously carried out repairs and the actual state) should be provided with a scientifically based methodology to identify priority areas for renovation and reconstruction according to the its actual state, including budget planning to maintain the reliability of the operation of technical facilities and systems at a given level. These basic imperatives determine the need for further development and distribution of the URRAN methodology—Resource and risk management based on analysis of reliability on railway transport. The use of the URRAN methodology provides an increase in the reliability and safety of the functioning of railway transport facilities. This methodology serves as the basis for creating system for risk and resource management at the stages of the life cycle of railway facilities. This risk and resource management system is based on an effective data collection system and intelligent information processing systems. Therefore, the priority task of the White Book is the development of the URRAN system by introducing blocks for automating the calculations of the remaining service life of infrastructure facilities, the formation of regulatory target budgets and the creation of a Unified corporate platform for managing resources, risks and reliability at the stages of the life cycle in railway domain."

Thus, the URRAN system of the Russian Railways has been entrusted with the key tasks of managing the technical maintenance of railway transport facilities based on the advanced RBM strategy—an asset management strategy based on risk assessment.

## 3.3 Object, Subject, Purpose, and Tasks of the URRAN System

### *3.3.1 Object, Subject, and Purpose of the URRAN System*

The URRAN system is being developed at the Russian Railways since 2010. Currently, URRAN is a set of regulatory and methodological support and hardware-software complex designed for integrated resource and process management in order to effectively provide railway transport services.

The object of URRAN application is the technical facilities, systems, and technological processes implemented by them.

The purpose of URRAN creation is to provide the adaptive management of a facility technical maintenance based on compliance with the criteria of reliability, safety, and economic efficiency of functioning at the stages of the life cycle, taking into account the risk assessment. Herein, by adaptive management we imply the form and methods of management (by enterprises) assuming the possibility and capability of the control system to change the parameters and structure of the regulator and the control subsystem in general depending on the change of internal parameters of the managed asset or the external environment (disturbances), as well as the changes in strategic goals. The purpose of use of the URRAN system is effective management of technical assets.

The purpose of URRAN introduction for railways is to increase the efficiency of railway transport operation based on the adaptive management of technical content under conditions of resource scarcity.

Each complex of JSC "RZD" facilities has specific features that are determined by the role of this complex in the transportation process and the conditions for the implementation of this role, as well as the established links with other complexes. Therefore, the goals of introducing the URRAN system are specific for each complex:

- Track facilities: Reduction of track infrastructure life cycle cost at the expense of resource redistribution under condition of ensuring the required level of operational reliability and acceptable level of train operation safety
- Signalling and remote control facilities: improving the operational reliability of signalling and remote control systems while ensuring an acceptable level of train delays and an acceptable life cycle cost based on the redistribution of resources
- Transenergo facility (electrification and power supply facility): the increase of electrification and power supply systems' life cycle based on risk assessment, under condition of ensuring the required level of operational reliability and acceptable level of train operation safety
- Telecommunication facility: Reduction of railway telecommunications' life cycle cost by increasing the efficiency of resource management based on improving the technology of telecommunication networks' operation under condition of ensuring the required safety performance and reliability when delivering telecommunication services
- Locomotive complex: Reduction of locomotive life cycle cost through the increase of efficient use of resources under condition of ensuring the required level of operational reliability and acceptable level of train operation safety
- Multiple units' rolling stock facility: Reduction of life cycle cost of multiple units' rolling stock facility through the efficient allocation of resources, under condition of ensuring the acceptable level of train operation safety while maintaining the requirements of passenger travel comfort.

### 3.3.2 *Tasks of the URRAN System*

To accomplish the said goals, the URRAN system is designed to solve the following tasks:

- Assess and predict the indicators of reliability and safety of infrastructure and rolling stock facilities in real time
- Manage technical and industrial risks
- Assess wear, remaining lifetime, and the limit state of railway transport facilities
- Predict the state of infrastructure facilities. Predict dangerous track failures
- Assess the life cycle costs of railway transport facilities

- Assess the activities of the Russian Railways divisions taking into account their results of work on ensuring the reliability and safety of the operated facilities
- Manage resources used for technical maintenance
- Provide decision-making support on the basis of a unified corporate platform URRAN.

When assessing and predicting the reliability and safety of railway transport facilities, it is necessary to take into account:

- The different nature of the tasks solved by the railway transport divisions. This implies a different physical maintenance and different units of measurement of facility lifetime spent. For example, the measurement unit of the operating lifetime for signalling and remote control facilities is the time while the measurement unit of the operating lifetime for the multi-unit rolling stock is wagon-km (section-km)
- The need to determine the current state of reliability and safety of facilities. For this purpose, the collection and processing of data on failures and technological violations should be carried out in real time
- Various structural and operational characteristics of facilities as well as various environmental conditions under which the facilities are operated. Therefore, in order to standardize, evaluate, and compare the indicators of operational reliability and safety of functioning of the track superstructure of various sections, a reference structures (object and elements) of facilities should be developed
- The difference between the operational reliability of the facility and the reliability of the service that the facility provides to the transportation process. For these different types of reliability, there are different reliability indicators and different approaches to ensuring reliability. So, to ensure the operational reliability of the facility, it is enough to create a rational system of its maintenance and repair, while to ensure the reliability of the service, this technical maintenance system is not enough. In this case organizational measures are also needed in order to reduce downtime in the transportation process.

The task of risk management within the URRAN system is aimed at solving the following issues:

- Ensuring the safety and reliability of the transportation process;
- Ensuring the occupational safety of activities related to technical maintenance of railway transport facilities;
- Ensuring fire safety;
- Balanced allocation of resources to ensure acceptable levels of safety of the transportation process and reliability of transport facilities.

The task of risk management within the URRAN system also includes minimizing the risks of pedestrian injury at pedestrian crossings.

When assessing wear, remaining service life and the limit state of railway transport facilities, it should be borne in mind that:

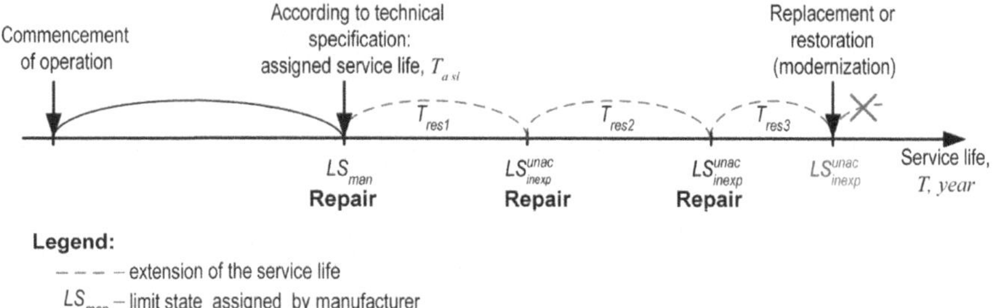


**Fig. 3.2** Time-base diagram of the transition from the assigned service life to the limit state of the facility

- The expired service life of facility does not mean its limit state in many cases. It is possible to carry out one or several restorations (repairs) of the facility after its service life assigned by the manufacturer (corresponds to the assigned service life specified in the technical documentation) was expired. As a result of said restorations (repairs), remaining service life is formed (Fig. 3.2) until the limit state is reached according to the criteria assigned by JSC Russian Railways in accordance with the URRAN methodology;
- The limit states of an object can be of only two types:
  - Limit state according to the criterion of unacceptability of further operation. The unacceptability of using a railway transport facility for its intended purpose is determined by non-compliance with the specified operational safety requirements, which is directly related to the excess of the acceptable level of risk due to the actual degradation of the technical condition of this facility due to the expiry of its service life during operation.
  - The limit state according to the criterion of economic inexpediency of operation. This criterion for making a decision on the limit state of a facility consists in the fact that at some point it is more profitable to replace the operated technical facility with a new one. This is because the extension of the service life results in increasing the costs of operation and maintenance and repair due to the physical wear of the facility.

- While ensuring the durability of railway equipment, it is necessary to distinguish between its physical and functional wear. The causes of physical wear are wearing-out, plastic deformation, fatigue, corrosion, and changes in the physical-chemical properties of structural materials. The causes for functional wear are fundamentally different and consist in the loss of facility value due to the use of new technologies and materials in the production of similar equipment. Functional wear is typical for high-tech facilities, such as signalling systems and communication facilities.

The task of predicting the state of infrastructure facility, in particular, pre-failures of the track, is one of the most costly and vital tasks of managing their technical maintenance. Because of the impact of the train load, over time, there accumulates faults in track level, which leads to a limitation of the train speed and the need to periodically perform work on the current track maintenance. The following should be done with the help of diagnostic complexes:

- Continuous monitoring and analysis of the evolution of the track state on the sections with different traffic intensity and length (railway, line, open lines, km, and railway milepost)
- Determination of the rate of track degradation
- Ranking of track sections (open line/station, km) in accordance with the risk matrix and assessment of the possible impact of the state of the track on the transportation process, depending on the traffic density
- Predicting the state of the track for a different time interval
- Forecasting the probable timing of occurrence of an undesirable and unacceptable state of the track
- Assessment of the state of the track section and the entire road
- Identification of track sections requiring repair
- Timely planning of repair activities in order to prevent the situation when the condition of a track may cause a failure
- Assessment of the efficiency of using the resources invested in the maintenance of the track (comparison of the track state before and after its repair), etc.

The task of estimating the life cycle cost of railway transport facilities covers all stages of facility design, development, manufacture, operation, etc., starting from the concept, setting requirements and ending with its disposal. Decisions on managing the cost of the life cycle of a facility are made at all stages of its life cycle and include:

1. At the stage of purchase: making decisions on purchasing new equipment (rails, points, multi-unit rolling stock, etc.), choosing a model and manufacturer
2. At the stage of operation:
   (a) Making decisions on options for organizing the technical maintenance and operation of the facility
   (b) Making decisions on economic feasibility and selecting projects when carrying out the modernization of the facility
   (c) Making decisions on determining the fact of facility wear
3. At the stage of completion of the life cycle: making decisions on determining the economic feasibility of extending the service life of the facility.

The task *of assessing the activities of the divisions of the Russian Railways*. In order to increase the impartiality of assessing the activities of divisions, when choosing methods and criteria for assessment, such basic factors as the state of technical equipment and the level of personnel competence are taken into account.

The following particular key indicators are used to assess the activities of railway transport divisions:

- Indicators for assessing the impact of the divisions activities on train traffic safety
- Indicators for assessing the impact of the divisions activities on the accuracy of timetable execution
- Indicators of the quality of the technical maintenance of the operated transport facilities.

The task of managing resources used for the technical maintenance of infrastructure facilities includes the following subtasks:

- The formation of principles for the classification of activities for modernization of the infrastructure of the Russian Railways for the purpose of its approval by public authorities
- Setting criteria for classifying infrastructure renewal activities
- Development of a model for determining dependencies that allows (on the basis of the acceptable URRAN indicators values) to determine the need for investments in the development and renewal of fixed assets as well as the amount of possible losses as a result of failures and unavailability of infrastructure
- Assessment of the impact of the amount of funds allocated for the development and renewal of infrastructure on safety and reliability indicators, as well as on the amount of possible losses as a result of failures and unavailability of the infrastructure
- Development of guidelines for proving to public authorities the need for financing the infrastructure renewal activities (overall repair and reconstruction of track sections).

The task of providing support for management decisions on the basis of a unified corporate platform URRAN is a large-scale work on informational support of the processes of:

- Collecting, analyzing, processing, and investigating events
- Decision-making by CEO, Railways, directorates, and the management of the Russian Railways.

Asset management does not focus directly on the asset itself, but on the value that the asset can provide to the organization. Asset management involves finding a balance between costs, opportunities, risks, and the required productivity of assets. Breaking this balance results in unreasonable costs or unacceptable risks of violating the safety of the transportation process. It is possible to achieve this balance if one implements Asset Ownership state-based Planning (AOP). It is neither more nor less than Results Based Management (RBM). This maintenance and repair strategy is based on the management of risk, revenues, and costs throughout the lifecycle of an asset. At the same time, effective investment decisions on the long-term period and investments in the development of technical personnel and information systems are expected.

The URRAN system is intended and implemented as AOP system deeply separated at the level of complexes of facilities, directorates, and structural divisions. It is enhanced with adaptive control technology, which expands the ability of the control system to change the parameters and even the structure of the regulator, depending on changes in the conditions of the transportation process.

## References

1. Iorsh, V.I.: Upravlenie aktivami v sootvetstvii so standartami ISO (Asset management in accordance with ISO standards). Upravlenie Kachestvom Journal. 5 (2015)
2. The Asset Management Landscape, 2nd edn. Available at: https://gfmam.org/sites/default/files/2019-05/GFMAMLandscape_SecondEdition_English.pdf
3. Koncepciya kompleksnogo upravleniya nadezhnost'yu, riskami, stoimost'yu zhiznennogo cikla na zheleznodorozhnom transporte (The concept of integrated management of reliability, risks, life cycle cost in rail transport), Moscow, RZD (2010)

# Chapter 4
# Basic Concepts and Indicators of Dependability and Functional Safety of Railway Transport Facilities

## 4.1 Basic Concepts and Definitions of Facility Dependability

*The facility* means any functional unit that can be considered separately (for example, a section of a railway track, a point, a section of a catenary system, a first class reliability relay, etc.). *The system* means a set of interrelated and interacting elements. *The device* is a technical means for performing one or several required functions. *Railway equipment* is a technical means or a set of technical means designed to ensure the transportation process in railway transport.

*Dependability* is one of the essential components of the quality of any technical facility. However, being only one of the quality elements, dependability differs significantly from all of its other properties, since it is the most general, complex property that characterizes the utility of any technical product, machine, device. Dependability is the only common property of the vast majority of industrial products. Dependability can be confirmed only over time. All other properties are determined by instantaneous values. Dependability is not subject to instrumental measurement, but is determined by calculations (probability calculations or statistical calculations) and testing of prototype models. The object of classical (structural) dependability [1] is a product (product: element, device, system). A product is an item or set of items manufactured in an enterprise. A product (technical facility) is the result of a production process. With the help of a technical facility or a set of facilities, a service (function) is provided. In other words, an action and, more generally, a process are performed.

*Services* are types of activities, works, in the process of performing of which a new, previously non-existent tangible product is not created, but the quality of an existing product is changed. The very provision of the service creates the desired result [2]. Service is the result of at least one action. The service has specific properties: intangibility, inseparability from its source (it is impossible to separate the process of providing information from the one who provides it). There are



I. B. Shubinsky, A. M. Zamyshlaev, *Technical Asset Management for Railway Transport*, International Series in Operations Research & Management Science 322,
https://doi.org/10.1007/978-3-030-90029-8_4

communication services and services of fulfilling functions, processes. There are informational and, more generally, technological processes.

*A technological process* is an ordered sequence of interrelated actions that are performed from the moment the initial data appears until the desired result is obtained. Thus, *the dependability of the technological process is the dependability of the service*. Evaluation and prediction of the dependability of a technological process is the subject of the theory of functional dependability of systems and processes [3].

The complexity of the "dependability" property is expressed in the fact that a technical device is more dependable, the less often it fails, the longer it remains in operable state, the easier and cheaper it recovers after a failure. For a more complete and detailed assessment, particular properties are introduced into consideration, which make up the integrated property "dependability": *reliability, durability, maintainability, and storability*.

*Reliability* is the property of a facility to continuously maintain an operable state for some time or some operating life under specific operating conditions. Operating life is the length of life during which a facility is in operation. *Operating life* can be either a continuous value (length of life during which a facility is in operation measured in hours, machine hours, kilometers, cubic meters of excavated soil, tons of handled cargo, kilometers driven, tonnage handled, etc.) or an integer value (number of working cycles, number of start-ups, etc.).

*Durability* is the property of a facility to maintain an operable state until the limit state provided that a system of maintenance and repairs is introduced.

*Maintainability* is a property of a facility to adapt to preventing, detecting causes of failures and damage as well as maintaining and restoring an operable state through maintenance and repairs. Maintainability is a complex property of dependability, determined by binding a facility to the specific conditions of its operation. A facility is maintainable if the following requirements are met: a facility is under maintenance, able to be recovered and is repairable. Facility maintenance means a set of operations to keep a facility in an operable state or to recover its operable state when using this facility for its intended purpose. Maintainability is characterized by checkability, availability, replaceability, modularity, changeability, and recoverability.

*Storability* is the property of a facility to keep within the specified limits the values of the parameters characterizing the ability of a facility to perform the required functions during and after its storage and (or) transportation.

*Availability* is a complex property of a facility to perform specific functions at any time.

Dependability is identified during the operation of a system and is assessed by quantitative indicators determined on the basis of processing experimental and statistical data.

Structural dependability is based on the concept of the state of an object.

*An operable state* is the state of a facility when it is able to perform all the functions in accordance with a technical specification provided that the necessary

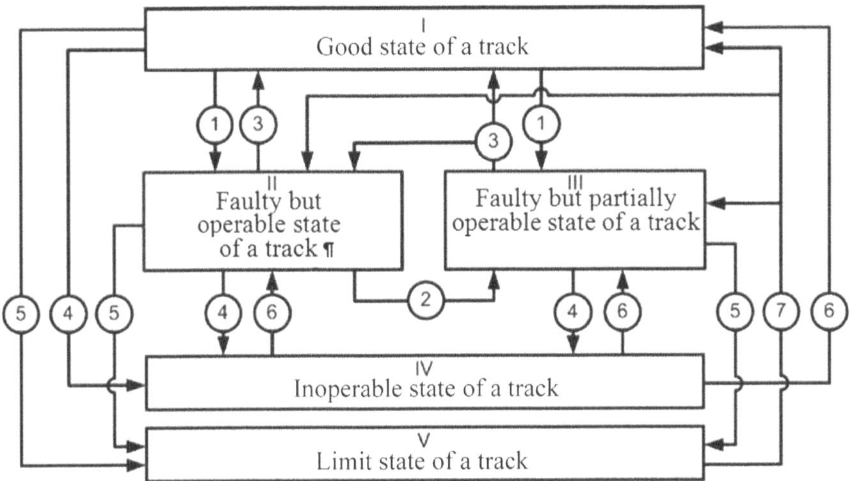


**Fig. 4.1** The diagram of the relations of the main track states: 1—accumulation of defects; 2—partial track failure; 3—current maintenance of the track; 4—complete track failure; 5—transition of the track to a limit state; 6—restoration of a track operable state; 7—overhaul, intermediate overhaul, or track raising

resources are provided (a facility may be in a operable state in terms of some functions and in inoperable state in terms of other functions at the same time).

*An inoperable state* is a state of a facility when it is unable to perform all the functions in accordance with a technical specification for any reason. *Downtime* is an inoperable state of a facility due to an internal reason. *An inoperable state due to external reason* is an inoperable state of a facility when it is unable to perform the required function due to the absence or lack of external resources. *An inoperable state due to internal reason* is an inoperable state of a facility when it is unable to perform a required function due to an internal malfunction or scheduled maintenance.

*A pre-failure state* of railway equipment is a faulty state of railway equipment when the probability of its transition to an inoperable state during a given operating life does not exceed the acceptable value.

*An accident* is an event caused by the transition of railway equipment to an inoperable state and/or a deviation from the specified modes of performing technological processes on the infrastructure of the Russian Railways, including due to external action, and resulting in a violation of the timetable.

*A good state* of a technical facility is such a state when it meets all the requirements of regulatory and technical documentation. If a technical facility does not meet at least one of these requirements, then it is in a faulty state. Being in a faulty state, a technical object can be operable or inoperable, depending on the level of malfunction influence on the performance of its functions in specific operating conditions. If a technical facility is in good state, then it is certainly operable.

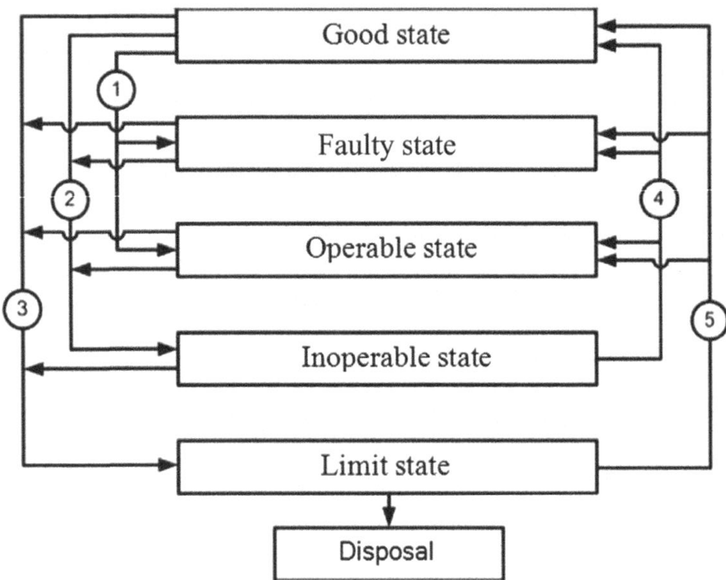


**Fig. 4.2** A standard diagram of the relations between the facility dependability states and the events of changing these states: 1—damage; 2—failure; 3—transition of the facility to a limit state; 4—recovery; 5—repair

Figure 4.1 illustrates the relations of facility states using the example of a track section.

Figure 4.2 shows a standard diagram of the relations between the dependability states of a facility and the events changing these states.

*A failure* of a technical facility is understood as an event as a result of which technical facility operability is lost or the established procedure for normal operation and repair is violated. *A failure* of a device or a part of a technical facility is an event due to which their operability is lost with the need for repair or replacement. *A defect* of a technical facility is a malfunction that does not violate the established procedure for normal operation and repair. The concept of "failure" applies not only to a technical facility, but also to the services that this technical facility provides while it is in operation. In this case, the term *service failure* is used. *Damage* is an event consisting in a violation of a facility good state while maintaining its operable state.

*Complete failure* is an event resulting in a complete loss of the facility operability. As for a facility that performs many functions, a complete failure in performing one or more functions means a partial failure of a facility as a whole.

*Partial failure* is a failure that results in a decrease in a facility differential output effect or not providing a given value of the integral output effect. Failure consequences are phenomena, processes, events, and states caused by the occurrence of a facility failure.

## 4.2 Features of Dependability Terminology in International Standards

The structural dependability (Dependability) of technical systems is considered as the "science of failures" in international publications [1]. It includes the properties of Reliability, Availability, Maintainability, and Safety. This means that the properties of durability and storability are taken out of the technical aspects of dependability and referred to the problem of the life cycle of a facility (system). At the same time, very important issues of analysis and synthesis of availability and functional safety (criticality or catastrophic failure) are referred to the dependability. The four listed properties that make up the structural dependability of technical systems are indicated as RAMS that is an acronym for Reliability, Availability, Maintainability, and Safety in English. Much attention is paid to the issues of providing RAMS.

Early in 2015, the International Electrotechnical Commission (IEC) adopted a new International Standard (IS) 60050-192 [4], which specifies the main terms in the field of dependability and gives their definitions. It forms Part 192 of the International Electrotechnical Vocabulary (IEV). This standard replaced IS 60050-191, adopted in 1990, as well as its amendments adopted in 1999 and 2002. Dependability terminology is given in relation to a (technical) item. In the MS 60050-191 standard, this term was defined simply by listing different types of objects: an individual part, component, device, subsystem, functional unit, equipment, or system that can be considered separately. However, this is hardly a complete list of all possible types of items. Therefore, in the new IS, an item is simply defined as a subject matter, and the list of these types of items was transferred to the note. Then the terms sub-item, system, and subsystem are defined.

Another note indicates that an item may consist of hardware, software, people, or any combination thereof. The terms "repaired/non-repaired item" used in the previous IS were replaced by more precise term "repairable/non-repairable item." The fact is that a word combination "repaired item" may be understood in two ways: as an item the repair of which is possible, or as an item the repair of which is being carried out at this moment. To exclude the second incorrect meaning the terms were replaced.

In IS 60050-191, the definition of the key term "dependability" actually reduced to enumeration of its properties: availability, reliability, maintainability, and maintenance support performance. The standard IS 60050-192 [4] gives the new definition of dependability: an ability of an item to perform as and when required. This definition has a note that specifies the properties of dependability. They are availability, reliability, recoverability, maintainability and maintenance and repair support performance, and in some cases also durability, safety, and security. Although safety and security are mentioned in the note as individual terms that have definitions, none of them is mentioned in IS 60050-192. The new term "recoverability" is defined as the ability of an item to recover from failure without repair. Indeed, often recovery is carried out, for example, by switching to backup or software reloading. These actions cannot be attributed to repair; therefore, the ability to such recovery is not covered by "maintainability," and it required the introduction of a new term. A particular case of recoverability is self-recoverability, when an item has the ability to recovery without external action to an item.

Speaking about the properties that are a part of dependability, we note that there is no item storability property in the IS. The new standard, as the previous IS, has a section dedicated to the items' states. The interstate standard GOST 27.002-15 defines two pairs of states: good–faulty, operable–inoperable (a good item is always in operable state, a faulty item can be both operable and inoperable; an operable item can be in good and faulty state, an inoperable item is always faulty). IS contains no equivalents to the good and faulty states, however, there are a number of other terms that specify the various states of an item. Particularly, there are operating and non-operating states. Being in the first state, the item performs any required function; being in the second state, it does not perform any required function. The time of being in this state is defined for each state. Then the times related to the maintenance and repair of an item are also determined. Terms of time include the useful life, as well as the early life failure period (infant mortality period), the constant failure intensity period, and wear-out failure period. The last three terms are typical for objects with a U failure rate curve. It should be noted that the U curve of the failure rate was described in the Russian literature since in the 1950s. However, there was no terminological support for it.

Some terms were excluded from the section about failures. For example, such types of failures as critical and non-critical, sudden and gradual, relevant and non-relevant, degradation, etc. At the same time, the following types of failures remained in the section are kept: complete and partial, primary and secondary, systematic and etc.; software failure was added.

The terms "failure cause," "failure mechanism," "common cause failures," and "common mode failures" are also kept. The first two of them are quite clear, let us give the definitions of the last two terms. Common cause failures are failures of various items resulting from a single cause: these failures would be considered independent of one another without considering the cause. Common mode failures are failures of various items, characterized by the same type of failure. A better understanding of this term is facilitated by the introduction of the term "failure mode," defined as manner in which failure occurs. The terms "failure effect" and "criticality" are also introduced. The failure effect is considered within and beyond the boundaries of the failed item. Criticality is the severity of the consequences in accordance with the established assessment criteria.

## 4.3 Basic Concepts and Definitions of Functional Safety of Facilities

Such notions as "safety" and "survivability" are important for facilities that are a potential source of danger.

*Safety* is the property of a facility not to pose a threat to the life and health of people, as well as to the environment while it is being manufactured and operated, or in case of its malfunction. Although safety is not a part of the notion "dependability," however, it is closely related to this notion under certain conditions, for example, if failures can lead to conditions that are harmful to people and the environment (excess of the maximum permitted value).

The notions "survivability" falls in between the notions "dependability" and "safety." Survivability is understood as:

- The property of a facility, consisting in its ability to withstand the development of critical failures resulting from defects and damage providing that there is an established maintenance and repair system, or
- The property of a facility to maintain limited operability under the influence of non-specified operating conditions, or
- The property of a facility to maintain limited performance in the presence of defects or damage of a certain type, as well as in case of failure of some components.

An example is the ability of structural elements to maintain sustaining capacity when fatigue cracks occur providing that its size do not exceed the specified values.

*Safety* of railway transport facilities is the ability to operate without accidents that lead to the 1st category failures. *A hazardous failure* is a facility failure of the first category. With regard to signalling and remote control systems, a hazardous failure is an event in which the operable and protective state of signalling and remote control systems is violated.

The basic concept of functional safety includes: *a safety-related system; safety state; safety function and safety integrity; safety integrity level.*

A safety-related system is a special system that:

- Ensures the performance of safety functions that are necessary to achieve or maintain a safe state of the facility controlled
- Designed to achieve the required safety integrity—alone or in conjunction with other safety-related embedded systems and external risk mitigation tools.

A safety-related system can be designed to:

(a) Prevent of a hazardous event (i.e., hazardous events do not occur when such a system performs its functions)
(b) Mitigate the consequences of a hazardous event by reducing the risk due to reducing the consequences
(c) Provide a combination of functions specified in subparagraphs (a) and (b).
(d) The software of these systems is also related to safety. Examples are software of control systems for aircraft, automobiles, interlocking and central traffic control systems on the railway, control system for nuclear reactors, etc. In other words, any software whose failure could result in the occurrence of an accident situation should be considered as safety-related software.

A person can be a part of a safety-related system. For example, a person can receive information from a programmable electronic device and perform safety actions on the basis of this information, or a person can perform safety actions using such a device.

***Safety states:***

1. *Good or faulty state*
2. *Operable or inoperable state*
3. *Protective state* is the state of the system in which the execution of all provided system functions is disabled in case of timely detection of a failure of any control element or ensuring the safety of the control. The protective state only occurs in safety-related systems. For an example, a protective state for two-channel safety-related system is formed by disconnecting the system from the controlled equipment and recovering the failed channel in case of detecting an inoperable state of at least one channel. The system functions are not performed in the protective state but all the intended safety functions are performed
4. *Hazardous state* is an inoperable state of a system in which at least one safety function is not performed
5. *Non-hazardous state* is an operable or protective state of the system.

It may happen that the pre-existing (safe) state of the system cannot be reached at a given time, for example, when the controlled facility is a running train. In this case, the functioning of the system can be disabled after a stop of the train.

Any control system or its part implements safety functions.

***A safety function*** is a function performed by a safety-related system or by external risk mitigation system and designed to provide or maintain a safe state in case of a specific hazardous event.

***Safety integrity*** is the level of satisfactory performance by a safety-related system of the required safety functions under all specified conditions for a specified period of time. The higher the safety integrity level of safety-related systems, the less likely these systems will fail while performing the required safety functions. The essence of safety integrity is as follows. All safety-related systems are divided into several groups in terms of qualitative characteristics and quantitative values. In the standards IEC 61508-12, IEC 62278, etc., four safety integrity levels (SIL) are given. Level 1 (SIL 1) is the lowest level, but it requires the use of great experience in development. Note that SIL 1 appears to in IEC 61508-12 standard [5] and other documents as "non-safety related." The second level (SIL 2) is not much more difficult to achieve than the previous one, but it requires more checks and tests to ensure it. This level requires a good design and application practice in accordance with of ISO 9001 requirements. As a result, the cost of the system increases. It is required more significant efforts and higher competence of developers than for the first and second levels to achieve the third and fourth safety levels (SIL 3 and SIL 4). System cost and time for its development are important factors. Table 4.1 gives the quantitative values of the system safety integrity levels, represented by the indicators of the rate of dangerous failures and the probability of a dangerous failure during an hour of operation.

**Table 4.1** Safety integrity levels (SIL)

| Safety integrity level (SIL) | High demand mode (dangerous failures per hour) $Q_D(1) = \lambda_D$ | Low demand mode (probability of failure) $Q_D(1) = \lambda_D$ |
|---|---|---|
| 4 | $\geq 10^{-9} - <10^{-8}$ | $\geq 10^{-5} - <10^{-4}$ |
| 3 | $\geq 10^{-8} - <10^{-7}$ | $\geq 10^{-4} - <10^{-3}$ |
| 2 | $\geq 10^{-7} - <10^{-6}$ | $\geq 10^{-3} - <10^{-2}$ |
| 1 | $\geq 10^{-6} - <10^{-5}$ | $\geq 10^{-2} - <10^{-1}$ |

Note that if we determine the values of SIL requirements at high demand mode, given in the middle column of the table, in other dimension (dangerous failures per year), then the middle and right columns would coincide numerically. Such a representation could cause confusion, since these are fundamentally different parameters, having even different dimensions. In this regard, in order to avoid this confusion, a different dimension was chosen (the number of failures per hour).

The reason that SIL requirements are actually specified in two tables (for high and low demand modes) is that both of the above approaches may be required to set the safety integrity level.

## 4.4 Basic List of Dependability and Functional Safety Indicators for Railway Transport Facilities

All indicators of structural dependability belong to one of two categories: quantitative indicators, qualitative indicators. A quantitative indicator is a "measure" of one or more properties of dependability. A qualitative indicator of dependability is the result of processing the opinions of a customer or a user, or an expert, or a developer, or their combined opinions. For example, a result of such processing may be such a result as "the dependability of the facility A is not worse than the dependability of the facility B" or "dependability of the facility A is acceptable." Qualitative indicators are often subjective. Therefore, preference is given mainly to quantitative indicators of dependability.

There are following dependability indicators: *single; integrated; calculated; experimental; operational.*

The value of the dependability indicator can be:

- Normative (in accordance with the regulatory documentation)
- Acceptable (it is determined on the basis of the current requirements for railway equipment depending on certain conditions of the transportation process and associated risks)
- Design (specified by the design (technical) documentation for the railway equipment).

Each of the dependability properties can be described quantitatively using some variable, the value of which characterizes the dependability of a facility with respect

**Table 4.2** Legend and definitions of indicators of facility reliability

| Indicator name | Legend | Dimension | Definition |
|---|---|---|---|
| Probability of failure-free operation | $P(x)$ | Dimensionless | The probability that the facility will not fail within a given operating life. |
| Probability of failure-free locomotive operation per trip | $P_1(x)$ | Dimensionless | The probability that a failure of a locomotive unit will not occur during a given operating life x (one trip) |
| Probability of failure | $Q(x)$ | Dimensionless | The probability that at least one failure of a certain type will occur within a given facility operating life. |
| Mean time to failure (for non-recoverable facilities) | $T_1$(MTTF) | Amount of operational work performed (time) | The mathematical expectation of the operating life of the facility until the first failure occurs. |
| Mean time between failures (for recoverable facilities) | $T_f$(MTBF) | Amount of operational work performed (time) | The ratio of the total operating time of the facility to the mathematical expectation of the number of its failures during this operating life |
| Failure rate | $(x)$ | Number of events per unit of time | The ratio of the mathematical expectation of the number of facility failures to its operating time within the observation interval |

to this property. This variable is called an indicator of a given dependability property or a single indicator of dependability. The measure of the facility dependability in regard to a set of several properties is called an integrated indicator of the facility (system) dependability. Let us consider the applied single and integrated indicators of the facility dependability. The following groups of single indicators of the dependability of railway transport facilities are defined below: reliability, maintainability, durability, and storability. In addition, there are integrated availability indicators.

**Reliability Indicators**

There is a basic list of reliability indicators for facilities and/or elements of all complexes of railway transport (see Table 4.2).

Dimensions for indicators of dependability and safety of railway transport facilities are given in Table 4.3.

**Availability Indicators**

Availability indicators are integrated indicators of the dependability of technical facility and represent a joint measure of their reliability and maintainability properties.

There is a basic list of availability indicators for facilities and/or elements of all complexes of railway transport (see Table 4.4).

**Durability Indicators**

The list of durability indicators for railway transport facilities is presented in Table 4.5.

**Table 4.3** Dimensions for indicators of dependability and safety

| No | Railway transport complex | Dimension |
|---|---|---|
| 1 | Track and structure facilities | Bln. t * km of ton-kilometer operation |
| 2 | Signalling and remote control facilities | Time (h) |
| 3 | Electrification and power supply facility | Mln. kW*h of transformed electrical energy (or h) |
| 4 | Telecommunication facility | Time (h) |
| 5 | Locomotive complex | Mln. locomotive*km of total run |
| 6 | Multiple unit rolling stock facility | Wagon*km (section*km) |

**Table 4.4** Legend and definitions of availability indicators for railway transport facilities

| Indicator name | Legend | Dimension | Definition |
|---|---|---|---|
| Availability coefficient | $K_A$ | Dimensionless | The probability that the facility will be in an operable state randomly at any time, except for the planned periods of time during which the use of the facility for its intended purpose is not provided |
| Operational availability coefficient | $K_{OA}$ | Dimensionless | The probability that the facility will be in an operable state randomly at any time, except for the planned periods of time during which its intended use is not provided, and, starting from this moment, it will operate without failures for a given time period |
| Technical utilization coefficient | $K_{TU}$ | Dimensionless | The ratio of the mathematical expectation of a total time during which the facility is in an operable state within some operating period to the mathematical expectation of the total time the facility is in an operable state and a downtime, due to scheduled maintenance (repair) and restoration, within the same period |
| Fleet availability | $A_f$ | Dimensionless | The ratio of the number of facilities that are in operable state to the total number of fleet facilities randomly at any time |

Durability indicators are necessary to determine the service life of track facilities, electrification and power supply facilities, locomotive complex facilities, and multiple unit rolling stock facilities. The application of these indicators to facilities of signalling and remote control, as well as to communications is limited only to those facility elements that fail gradually (for example, signalling racks, cables, relays, etc.)

When using indicators of durability, the starting point and the type of action after reaching the limit state should be indicated (for example, the gamma percentage life from the second overhaul to write-off).

The average life (operating life) is estimated based on the results of monitoring several facilities of the same type. The life of each specific facility depends on many

**Table 4.5** Legend and definitions of durability indicators for railway transport facilities

| Indicator name | Legend | Dimension | Definition |
|---|---|---|---|
| Average life | $\overline{X}_L$ | Amount of operational work performed | Mathematical expectation of life |
| Gamma-percentile life | $\overline{X}_{GP}$ | % | The total operating life during which a facility does not reach the limit state with the probability γ, expressed as a percentage |
| Average service life | $\bar{t}_{SL}$ | Time | Mathematical expectation of facility service life |
| Gamma percentage service life | $\bar{t}_{GPS}$ | % | Calendar duration of operation, during which a facility will not reach the limit state with a probability γ, expressed as a percentage |

**Table 4.6** Legend and definitions of storability indicators for reference railway transport facilities

| Indicator name | Legend | Dimension | Definition |
|---|---|---|---|
| Average storability time | $\bar{t}_S$ | Year | Mathematical expectation of storability time |
| Gamma percentage storability time | $\bar{t}_{GPS}$ | % | The storability time that a facility reaches with a given probability $\gamma$, expressed as a percentage |

factors; however, the criterion of the limit state is such values of the indicators monitored when the further operation of the facility is unsafe for humans and the environment or economically unprofitable.

A statistical estimation of the average life (service life) can be obtained based on the results of observation of n facilities of the same type, operated in approximately the same conditions. The life (service life) of each specific facility observed depends on many random factors, while the limit state of a facility is practically determined by its characteristics, indicating that its further operation becomes unsafe for humans and the environment, or becomes economically unprofitable.

**Storability Indicators**

Table 4.6 shows the list of storability indicators for objects and elements of track facilities. Storability indicators are necessary to assess the condition and the possibility of further use of machinery and equipment that are in a state of conservation or storage in specially equipped rooms.

To calculate the amount of the machines, devices, and equipment fleet for its replenishment, gamma percentage indicators at $\gamma = 50\%$ may be required, which approximately corresponds to the average values.

Storability indicators are introduced for track machines, for equipment of traction substations, and for rolling stock facilities.

**Safety Indicators**

Legend and definitions of safety indicators for railway transport facilities are given in Table 4.7.

**Table 4.7** Legend and definitions of safety indicators for reference objects and elements

| Indicator name | Legend | Dimension | Definition |
|---|---|---|---|
| Probability of safe operation | $P_{SO}(x)$ | Dimensionless | The probability that a dangerous failure of a facility will not occur within a given operating life |
| Dangerous failure probability | $Q_{D}(x)$ | Dimensionless | The probability that at least one dangerous failure of a facility will occur within a given operating time |
| Mean time to dangerous failure (for non-recoverable facilities) | $T_{DI}$ | Amount of operational work performed (time) | The mathematical expectation of the operating life of a facility until the first dangerous failure occurs |
| Mean time between dangerous failure (for recoverable facilities) | $T_{DF}$ | Amount of operational work performed (time) | The ratio of the total operating life of the recoverable facility to the mathematical expectation of the number of its dangerous failures during this operating life |
| Mean time between protective failures | $T_{P}$ | Amount of operational work performed (time) | The ratio of the total operating life of a recoverable facility to the mathematical expectation of the number of its dangerous failures during this operating life |
| Dangerous failure rate | $\lambda_{D}(x)$ | Number of events per unit of time | The ratio of the number of dangerous failures for the operating life spent of the facility to its operating life |
| Mean time to return to a safe state | $\bar{t}_{RSS}$ | Time | Mathematical expectation of the time for elimination of a dangerous failure of a facility |
| Safety coefficient | $K_{S}$ | Dimensionless | The probability that a facility will be in a safe state randomly at any time |

$x$ facility operating life

## 4.5 Specific Maintainability Indicators of Facilities of Infrastructure and Rolling Stock Complexes

### Track Facilities

*A complete list of maintainability indicators* for the facilities of track repair complex is given in [6]. In this book, we will limit to a list of operational indicators of the maintainability of the track superstructure (see Table 4.8).

*Calculation of maintainability indicators* is carried out in accordance with the formulas given in [1, 6]. The calculation of the average downtime, as well as indicators for assessing the recoverability of track facilities is based on a set of factors shown in Fig. 4.3.

**Table 4.8** A list of maintainability indicators for the track superstructure

| Indicator name | Legend | Dimension |
|---|---|---|
| Average downtime | $T_{DT}$(MDT) | h |
| Average time to recovery | $T_R$(MTTR) | h |
| Average operating life between scheduled preventive maintenance of a given type | $\overline{X}_{SPM}$ | Million tons gross |
| Average operating life between scheduled repairs of a given type | $\overline{X}_{SR}$ | Million tons gross |
| Average duration of maintenance of a given type | $\tau_{DM}$ | h |
| Average duration of repair of a given type | $\tau_{DR}$ | h |
| Total duration of maintenance | $\tau_{DM}^{\Sigma}$ | h |
| Total duration of repair | $\tau_{DR}^{\Sigma}$ | h |
| Specific total duration of maintenance | $\tau_{SDM}^{\Sigma}$ | h/million tons gross |
| Specific total duration of repair | $\tau_{SDR}^{\Sigma}$ | h/million tons gross |
| Average labor intensity of maintenance of a given type | $S_M$ | Person-hour |
| Average labor intensity of repair of a given type | $S_R$ | Person-hour |
| Total labor intensity of maintenance | $S_M^{\Sigma}$ | Person-hour |
| Total labor intensity of repair | $S_R^{\Sigma}$ | Person-hour |
| Specific total labor intensity of maintenance | $S_{MS}^{\Sigma}$ | Person-hour/million tons gross |
| Specific total labor intensity of repair | $S_{RS}^{\Sigma}$ | Person-hour/million tons gross |

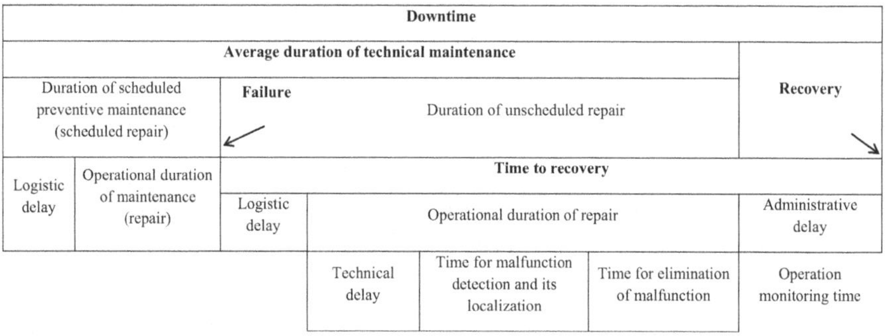

**Fig. 4.3** Chart of the summands of the duration of facility operability and inoperability

### Complexes of Signalling and Remote Control Facilities, and Electrification and Power Supply Facilities

The following main indicators (Table 4.9) are used for these complexes when managing the maintainability of its facilities.

### Telecommunication Complex

*The main maintainability indicators*, developed for facilities of the communications complex are shown in Table 4.10.

**Table 4.9** Main maintainability indicators for signalling and remote control complex and electrification and power supply complex

| Indicator name | Legend |
|---|---|
| Average duration of scheduled maintenance (repair) of a given type | $\bar{t}_{SM}(\bar{t}_{SR})$ |
| Average time (duration) to recovery | $\bar{t}_R$ |
| Average operational duration of scheduled maintenance (repair) of a given type | $\bar{t}_{OSM}(\bar{t}_{OSR})$ |
| Average operational duration of recovery | $\bar{t}_{OM}$ |
| Average labor intensity of recovery | $\overline{S}_R$ |
| Average cost of recovery | $\overline{C}_R$ |
| Average operating life between scheduled types of maintenance (repair) 1 | $\overline{X}_M(\overline{X}_R)$ |
| Average labor intensity of scheduled maintenance (repair) of a given type | $\overline{S}_M(\overline{S}_R)$ |
| Average cost of scheduled maintenance (repair) of a given type | $\overline{C}_M(\overline{C}_R)$ |

**Table 4.10** Maintainability indicators for facilities of the communications complex

| | Indicator name | Legend |
|---|---|---|
| 1. | Average time to eliminate an accident for a facility (complex facility) or a group of facilities (complex facilities) of one type | $\bar{t}_A$ |
| 2. | Average labor intensity of an accident elimination for a facility (complex facility) or a group of facilities (complex facilities) of one type | $\overline{S}_A$ |
| 3. | Average time to service recovery for a complex facility "communication network" or a group of complex facilities "communication network" of one type | $\bar{t}_{SR}$ |
| 4. | Average labor intensity of service recovery for a complex facility "communication network" or a group of complex facilities "communication network" of one type | $\overline{S}_{SR}$ |
| 5. | Average time to a recovery of a facility (complex facility) or a group of facilities (complex facilities) of one type | $\bar{t}_R$ |
| 6. | Average labor intensity of recovery of a facility (complex facility) or a group of facilities (complex facilities) of one type | $\overline{S}_R$ |
| 7. | Average time of non-provision of services due to scheduled preventive maintenance (scheduled repairs) for a complex facility "communication network" or a group of complex facilities "communication network" of one type | $\bar{\bar{t}}_{SMP}\left(\bar{\bar{t}}_{SPR}\right)$ |
| 8. | Average duration of scheduled preventive maintenance (scheduled repair) of a facility (complex facility) or a group of facilities (complex facilities) of one type | $\bar{t}_M(\bar{t}_R)$ |
| 9. | Average labor intensity of scheduled preventive maintenance (scheduled repair) of a facility (complex facility) or a group of facilities (complex facilities) of one type | $\overline{S}_M(\overline{S}_R)$ |

*The calculation of the maintainability indicators* for the facilities of the telecommunication complex is carried out in accordance with the formulas given in [7].

**Locomotive Complex**

According to Fig. 4.3, it takes the time indicated below for the recovery of the ith locomotive (wagon, motrice, etc.) after its jth failure

$$t_{\mathrm{R}ij} = t_{\mathrm{UR}ij} + t_{\mathrm{AUR}ij} = t_{\mathrm{LUR}ij} + t_{\mathrm{OUR}ij} + t_{\mathrm{AUR}ij}$$

where:

$t_{\mathrm{R}ij}$ is the time to the recovery of *i*th locomotive after its *j*th failure during the observation interval, h

$t_{\mathrm{UR}ij}$ is the duration of unscheduled repair of the *i*th locomotive after its *j*th failure during the observation interval, h

$t_{\mathrm{AUR}ij}$ is the administrative delay during unscheduled repair of the *i*th locomotive after its *j*th failure during the observation interval, h

$t_{\mathrm{LUR}ij}$ is the logistical delay during unscheduled repair of the *i*th locomotive after its *j*th failure during the observation interval, h

$t_{\mathrm{OUR}ij}$ is the operational duration of unscheduled repair of the *i*th locomotive after the *j*th failure during the observation interval, h.

The duration of the *k*th scheduled preventive maintenance is determined as the sum of the following two components

$$\bar{t}_{\mathrm{Mk}} = \bar{t}_{\mathrm{LMk}} + \bar{t}_{\mathrm{OMk}}$$

where:

$\bar{t}_{\mathrm{Mk}}$ is the duration of the *k*th scheduled preventive maintenance of a given type locomotive, h

$\bar{t}_{\mathrm{LMk}}$ is the logistic delay during *k*th scheduled preventive maintenance, h

$\bar{t}_{\mathrm{OMk}}$ is the operational duration of the *k*th scheduled preventive maintenance, h.

The duration of the *j*th scheduled repair of a given type locomotive includes:

$$t_{\mathrm{R}j} = t_{\mathrm{LR}j} + t_{\mathrm{OR}j}$$

where:

$t_{\mathrm{R}j}$ is the duration of the *j*th scheduled repair of a given type locomotive, h

$t_{\mathrm{LR}j}$ is the logistic delay during the *j*th scheduled repair of a given type locomotive, h

$t_{\mathrm{OR}j}$ is the operational duration of the *j*th scheduled repair of a given type locomotive, h.

Management of the locomotive maintainability is carried out by the service organization. Therefore, maintainability indicators are associated with the special

**Table 4.11** Name, legend, dimension, and definition of key indicators of locomotive maintainability

| Indicator name | Legend | Dimension | Definition |
|---|---|---|---|
| Average time to recovery | $\bar{t}_R$ | h | Average time interval from the moment of locomotive failure to the moment of its operable state recovery. |
| Specific total labor intensity of scheduled preventive maintenance | $S_M^\Sigma$ | person-hour/(units of measurement of operating life) | The ratio of the average total labor intensity of the scheduled preventive maintenance of the locomotive to its average operating life over the observation interval. |
| Specific total labor intensity of scheduled repair | $S_R^\Sigma$ | person-hour/(units of measurement of operating life) | The ratio of the average total labor intensity of a scheduled locomotive repair to its average operating life over the observation interval. |

aspects of the organization of service work. At the same time, there are key indicators of locomotive maintainability that are common for various service organizations (see Table 4.11).

**Multiple Unit Rolling Stock**

The summands of time for the technical maintenance of multiple unit rolling stock facilities and the locomotive facilities are similar. At the same time, many facilities of the multiple unit rolling stock, such as electric multiple unit, motrices, diesel electric multiple unit, etc., differ significantly from locomotives in terms of maintainability characteristics and therefore are evaluated by a number of other indicators of maintainability. The main ones are shown in Table 4.12.

The calculation of the maintainability indicators for the facilities of the multi-unit rolling stock is carried out in accordance with the formulas given in [8].

**Table 4.12** Name, legend, dimension, and definition of maintainability indicators for multiple unit rolling stock (MU)

| Indicator name | Legend | Dimension | Definition |
|---|---|---|---|
| Average time to recovery | $\bar{t}_R$ | h | The average time interval from the moment of a MU failure to the moment of its operable state recovery |
| Recovery rate | $\mu$ | 1/h | The number of unscheduled repairs of the unit of MU per unit of time. |
| Average total time to recovery | $\bar{t}_R^\Sigma$ | h | The ratio of a sum of time intervals from the moment of MU unit failure to the moment of its operable state recovery to the number of units of MU. |
| Specific total labor intensity of unscheduled repair | $S_{NR}^\Sigma$ | Person-hour/(unit of operating life) | The ratio of the average total labor intensity of the unscheduled repair of an MU unit to its average operating life over the observation interval. |
| Average duration of maintenance | $\bar{t}_M$ | h | The average duration of one scheduled preventive maintenance of a given type of MU unit |
| Average duration of repair | $\bar{t}_R$ | h | The average duration of one scheduled repair of a given type of MU unit. |
| Specific total labor intensity of scheduled preventive maintenance | $S_M^\Sigma$ | Person-hour/(unit of operating life) | The ratio of the average total labor intensity of scheduled preventive maintenance of a MU unit to its average operating life over the observation interval. |
| Specific total labor intensity of scheduled repair | $S_R^\Sigma$ | Person-hour/(unit of operating life) | The ratio of the average total labor intensity of the scheduled repair of a MU unit to its average operating life over the observation interval. |
| Average duration of downtime | $T_{DT}$ | h | The average time interval during which a MU unit is down due to its recovery, preventive maintenance, and scheduled repairs over the observation interval |

# References

1. Shubinsky, I.B.: Strukturnaya nadezhnost' informacionnyh sistem. Metody analiza (Structural dependability of information systems. Methods of analysis). LLC "Journal Dependability", Moscow (2012)
2. Borisov, A.B.: Bol'shoj ekonomicheskij slovar' (Big Dictionary of Economics). Knizhnyj Mir, Moscow (2003)
3. Shubinsky, I.B.: Funkcional'naya nadezhnost' informacionnyh sistem. Metody analiza (Functional dependability of information systems. Methods of analysis). LLC "Journal Dependability", Moscow (2012)
4. IEC 60050-192:2015 International Electrotechnical Vocabulary (IEV) – Part 192: Dependability

5. GOST R IEC 61508-1-2012 Funkcional'naya bezopasnost' sistem elektricheskih, elektronnyh, programmiruemyh elektronnyh, svyazannyh s bezopasnost'yu (Functional safety of electrical, electronic, programmable electronic safety-related systems)
6. Metodika ocenki deyatel'nosti strukturnyh podrazdelenij remontnogo kompleksa putevogo hozyajstva OAO "RZD" po pokazatelyam nadezhnosti i bezopasnosti funkcionirovaniya zheleznodorozhnogo puti i putevyh mashin (Method for assessing the activity of structural divisions of the repair complex of track facilities of JSC "Russian Railways" in terms of dependability and safety indicators of the railway track and track machines functioning) Moscow, RZD (2012)
7. Metodika rascheta pokazatelej nadezhnosti zheleznodorozhnoj elektrosvyazi (Method for calculating dependability indicators of railway telecommunications) Moscow, RZD (2014)
8. Metodika rascheta pokazatelej nadezhnosti i bezopasnosti funkcionirovaniya motorvagonnogo podvizhnogo sostava (Method for calculating dependability and safety indicators of the operation of multiple unit rolling stock), Moscow, RZD (2014)

# Chapter 5
# Standardization of the Facilities of Railway Transport and the Normalization of Dependability Indicators

## 5.1 General Provisions

The facilities of railway transport are operated ***in various environmental conditions*** (temperature, humidity, wind, pressure, amount of precipitation, snow, floods, etc.), ***under various operating conditions*** (track layout, track class, set speed, etc.) and in accordance with ***different management approaches***. They are operated ***at different utilization rate***. The facilities of the same functionality may have ***significant design differences***. All these circumstances mean that facilities of the same functionality can fail with different frequency (intensity). Moreover, differences in failure rate can significantly vary depending on operating conditions. It follows that ***the dependability of a facility could not be evaluated by the total number of its failures at various sections of the railway network***.

This problem can be solved by introducing a system of coefficients for the operating conditions and the facilities design, i.e. by converting this facility to some reference facility. This will allow to compare the results of operation of railway transport facilities by structural divisions; to generalize statistical data on dependability and safety of facilities throughout the entire railway network; to identify the most problematic facilities in terms of dependability and safety of its design; to develop recommendations on improving the facility operating conditions, etc.

Object-element structures of infrastructure facilities and rolling stock have been developed to assess and compare the indicators of operational dependability and safety of functioning of infrastructure facilities and rolling stock for each complex of infrastructure and rolling stock. Object-element structures of infrastructure facilities and rolling stock are given below.



I. B. Shubinsky, A. M. Zamyshlaev, *Technical Asset Management for Railway Transport*, International Series in Operations Research & Management Science 322,
https://doi.org/10.1007/978-3-030-90029-8_5

**Table 5.1** Main characteristics of reference facilities of the railway track

| Facility | Reference object | Main characteristics of a reference object |
|---|---|---|
| Linear construction of a track | Reference kilometer of track superstructure | 1 km of continuous track with R65 rails, concrete sleepers, anchor fasteners, on broken stone ballast through which 100–200 million tons (gross) passed after its laying, operated in a straight section of the track at zero level in a temperate climate |
| Crossover track and crossings | Reference switch | R65 type switch, grade 1/11 with concrete beam on broken stone ballast |
| Roadbed | Reference section of roadbed | A section of roadbed the operation of which is not associated with complicated engineering—geological and climatic conditions; and this section was in operation for no more than 50 years with a traffic load of 50–80 million tons (gross)*km/km per year. |

## 5.2 Standardization of Infrastructure Objects

### *5.2.1 Track Complex*

**Reference Object-Element Structure of the Railway Track Superstructure**
The railway track forms the basis of railway transport and is a complex multi-element set of engineering structures and devices that form a road with a guiding rail gauge, designed to provide movements of trains. Generally, because of the impact of the train load, there accumulates faults in track level along the length of the track nonuniformly over time (track sagging, distortion of track, track being out of surface, track directional errors, etc.). It results in limitation of train speed and the need to periodically perform track current maintenance.

Assessment of technical condition of a railway track of a particular section in terms of operational dependability and safety indicators becomes complicated due to a large variety of designs types of structural elements operated under different operating and environment conditions. Thus, the following reference facilities [1, 2] (see Table 5.1, Figs. 5.1, 5.2 and 5.3) were adopted to standardize, assess, and compare the indicators of operational dependability and safety of operation of a railway track of various sections.

The calculation of dependability and safety indicators for the operation of reference objects of the railway track is presented in [1].

The conversion of the operated elements of the superstructure of the railway track (for each of the types) to the reference ones is made according to the following formulas:

**Fig. 5.1** Linear construction of a track with R65 rails, concrete sleepers, anchor fasteners, on broken stone ballast

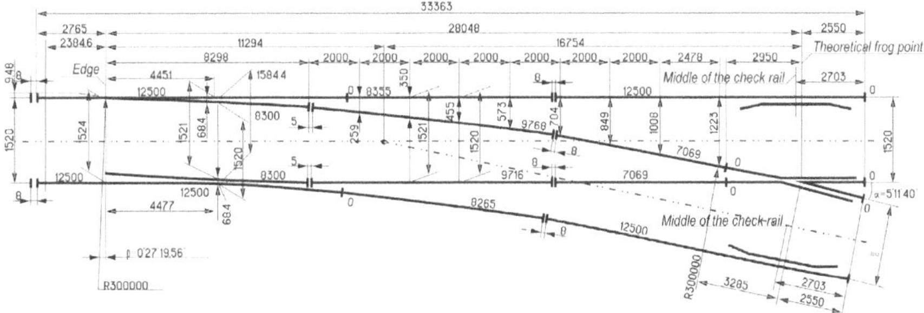


**Fig. 5.2** R65 type switch, grade 1/11, concrete beams

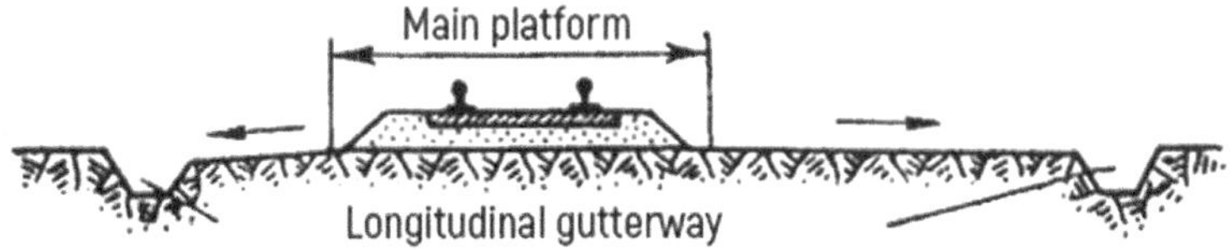


**Fig. 5.3** Roadbed with zero cross-section

- for a linear construction of a track:

$$N^{\mathrm{rTr}}(T_n) = L^{\mathrm{Tr}} \times k^{\mathrm{S}} \times k^{\mathrm{O}} \times k_i^{\mathrm{Cl}} \times k^{\mathrm{T}}(T_n)\ (\mathrm{pcs.}), \tag{5.1}$$

- for a switch:

$$N^{\mathrm{rSw}}(T_n) = L^{\mathrm{Sw}} \times k^{\mathrm{S}} \times k^{\mathrm{O}} \times k^{\mathrm{Cl}} \times k^{\mathrm{Sw}} \times k^{\mathrm{C}} \times k^{\mathrm{T}}(T_n)\ (\mathrm{pcs.}), \tag{5.2}$$

where:

$L^{\mathrm{Tr}}$ ($L^{\mathrm{Sw}}$) is the length of elements of the same design, operating under the same conditions, with same amount of operational work performed at a given polygon

$k^{\mathrm{S}}$ is the coefficient taking into account the structural features of the element (type of rails, type of ballast, sleepers spacing, switch grade, etc.)

$k^{\mathrm{O}}$ is the coefficient taking into account the operating conditions of the element (track class, set speed, and track layout)

$k^{\mathrm{Cl}}$ is the coefficient taking into account climatic conditions

$k^{\mathrm{Sw}}$ is the coefficient taking into account the number of switch movements

$k^{\mathrm{C}}$ is the coefficient taking into account the possibility of control of a switch (centralized/non-centralized)

$k^{\mathrm{T}}(T_n)$ is the coefficient taking into account the tonnage handled since the element construction or overhaul

$T_n$ is the volume of work performed (million gross tons) in the observation interval $n$.

The tables [1] give the values of the conversion coefficients for each of the types of elements, developed taking into account standard labor costs of personnel involved in the track current maintenance.

**Track Machines Are an Integral Part of the Track Complex**

According to their purpose, the track machines are divided into groups: for repair of the roadbed (spreader and cleaning machines); for ballasting and lifting of the track (electric ballasters, track lifter, leveling machine, and ballast hopper); for transportation and unloading of materials (hopper); for cleaning ballast (crushed stone cleaning machines); for laying of track and laying-in a switch (track-laying machines and rail layer); for welding and grinding of rails (machines PRSM, RShP-48); for track panel assembly site (track panel assembly machine and track panel dismantling machine); for track alignment, ballast tamping, and stabilization of the ballast layer (leveling and tamping machine, liner, and finishing machine); for diagnostics of track state (track measuring car, rail detector car, motorailer, and railcar); for snow cleaning (snow plow, rotary snow plow, and snow plough), as well as transport, traction, and power and loading-and-unloading facilities for engineering works (muck disposal train, hopper cars, motorailer, hand car, and autonomous car). The conversion of the operated track machines (for each of the type) to the reference ones is made according to the following formula:

$$N_i^{\mathrm{rTM}} = N_i^{\mathrm{TM}} \times k_i^{\mathrm{Cap}} \tag{5.3}$$

where:

$N^{\mathrm{TM}}$ is the number of track machines of the same design

$k^{\mathrm{Cap}}$ is the coefficient taking into account the capacity of the track machine.

The values of conversion coefficient for each type of track machines are summarized in tables [3].

### 5.2.2 *Reference Object-Element Structure of Railway Signalling and Remote Control Facilities*

Devices and systems of railway signalling and remote control facilities (hereinafter—SRCF) are the most important elements of the technical equipment of railway transport. These devices and systems ensure the continuity and safety of train operation and the increase in capacity of railway lines. The dependability of SRCF devices and systems operation largely determines the efficiency of using other technical means, especially rolling stock. This contributes to an increase in labor productivity and decrease in cost of transportation.

There are devices and systems used in railway transport for open lines (see Fig. 5.4) and for stations (see Fig. 5.5). Hence, it is necessary to consider the reference object of the SRCF for open lines and the reference object of SRCF for

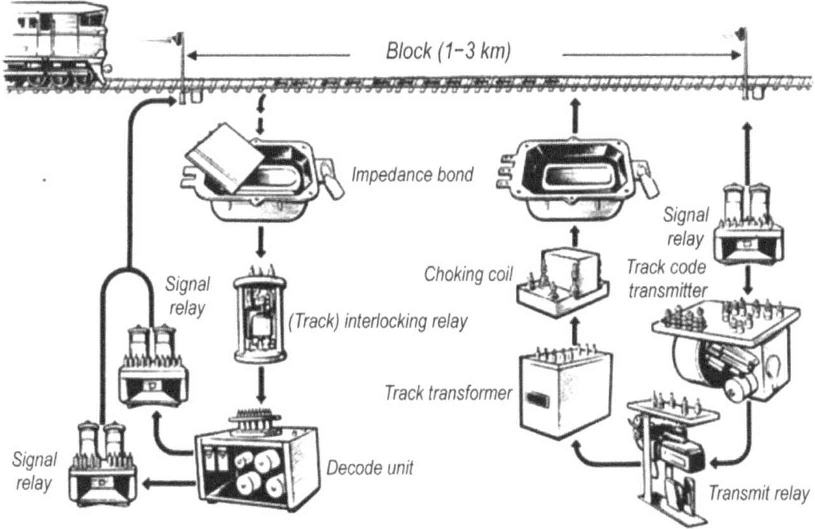


**Fig. 5.4** SRCF for open lines (track circuit with code automatic blocking system)

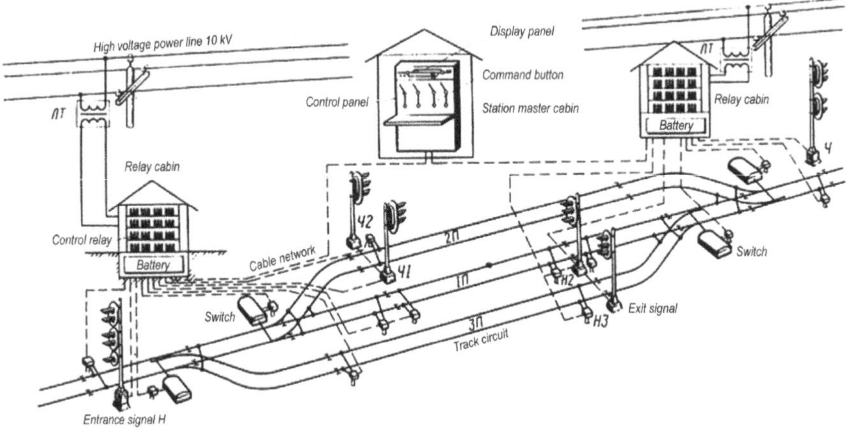


**Fig. 5.5** SRCF for stations (marshrutno-relay centralization of the station)

**Table 5.2** The main characteristics of the reference objects of SRCF

| Facility | Reference object | Main characteristics of a reference object | Reference object elements |
|---|---|---|---|
| SRCF for open lines | Reference block | DC electric traction<br>Automatic block system AB-2-K-25-50-ET-82<br>Signal current frequency 50 Hz;<br>Three aspect signalling<br>Equipment with central power supply | Occupation control device<br>Device for signalling and control of light signal<br>Device for reverse traction current sewerage (drainage)<br>Device for transmitting signals to a locomotive |
| SRCF for stations | Reference complex of technical means for controlling a switch | Switch section<br>DC electric traction<br>Station rail circuits with a frequency of 50 Hz with a two-element plug relay<br>Computer-based interlocking mrts-13<br>One electric switch machine with a direct current electric motor<br>The control chart of electric switch machine (double wire)<br>One 4-aspect ground light signal installed<br>Equipment with central power supply | Device for providing logical dependence<br>A set of switch and locking devices<br>Command and control device<br>Power supply device |

station [2, 4] (see Table 5.2). The reference object of SRCF for open lines is a reference block. It is a railway section in open lines, equipped with a numerical coded current automatic block system with direct current electric traction. Only one mobile unit may occupy a block at a time in accordance with the safety conditions of train traffic.

The reference object of SRCF for stations is a reference complex of technical means for controlling the switch. It is a railway section at the station, equipped with a switch connected to relay interlocking systems with one alternating current drive. Only one mobile unit may occupy a railway section at a time in accordance with the safety conditions of train traffic.

It is necessary to introduce conversion coefficient to convert facilities of a signalling and remote control systems to the reference ones. *Climatic conversion coefficients ($k^{\mathrm{Cl}}$)* is the coefficient of change in the intensity of the flow of failures of technical means depending on climatic conditions (effect of heat, cold, low (up to deep vacuum) pressure, humidity, wind, atmospheric pollutants, and aggressive environment). *Conversion coefficients for the section loading ($k^{\mathrm{L}}$)* is the coefficient of change in the intensity of failure flow depending on the intensity of train traffic along the section (station) (section loading is characterized by the line category). *Conversion coefficients for sections technical equipping ($k^{\mathrm{TE}}$)* is the ratio of failure flow rate of the actual technical equipment of the block (complex of technical means for controlling a switch) to the failure flow rate of the reference block (complex of technical means for controlling a switch).

It is possible to convert the operated signalling and remote control facilities of each type to the reference ones according to the following formulas:

- for the reference block:

$$N^{\mathrm{rB}} = L^{\mathrm{B}} \times k^{\mathrm{Cl}} \times k^{\mathrm{L}} \times k^{\mathrm{TE}} \tag{5.4}$$

- for the reference complex of technical means for controlling a switch:

$$N^{\mathrm{rSw}} = L^{\mathrm{B}} \times k^{\mathrm{Cl}} \times k^{\mathrm{L}} \times k^{\mathrm{TE}} \tag{5.5}$$

where:

$N^{\mathrm{rB}}$ is the length of blocks of the same design, operating under the same conditions at a given polygon

$N^{\mathrm{rSw}}$is the number of complexes of switch controlling technical means of the same design, operating under the same conditions at a given polygon.

### 5.2.3 *Reference Object-Element Structure of Railway Power Supply*

The main task of the railway electrification and power supply facilities is to maintain uninterrupted and high-quality supply of electric traction power to trains and all other railway consumers related, primarily, to movement of trains. In addition, railway electrification and power supply facilities provide non-traction consumers with electricity. Electrified lines, non-traction railway consumers, and non-transport consumers are supplied with electricity from regional substations and high-voltage power lines of power system operators of the Russian Federation through traction and transformer substations (see Figs. 5.6 and 5.7)

The following reference object-element structure given in Table 5.3 is adopted for normalization, assessment, and comparison of operational dependability and safety indicators for railway power supply facilities operation on various sections of railways.

The calculation of dependability and safety indicators for operation of reference objects of railway power supply is presented in [5].

The conversion of the operated electrification and power supply facilities (for each of the types) to the reference ones is made according to the following formulas:

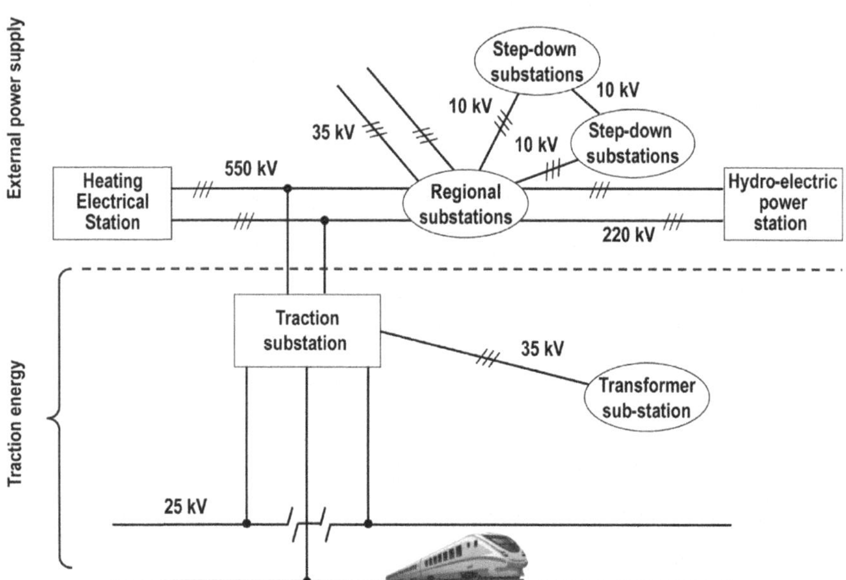


**Fig. 5.6** Schematic diagram of railway power supply

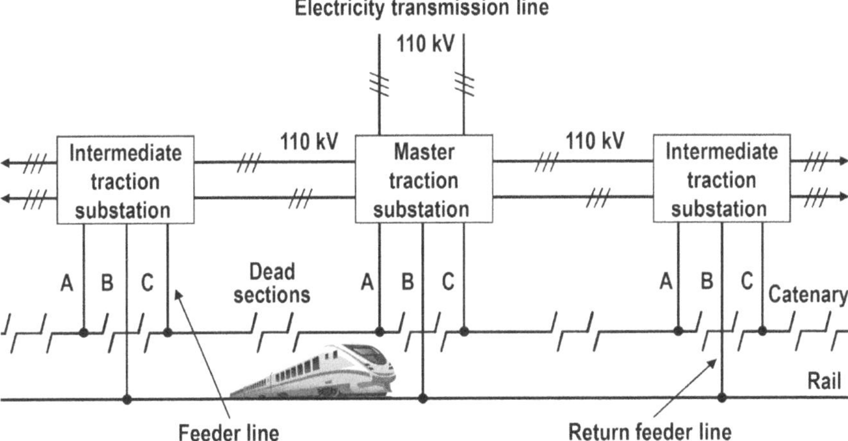


**Fig. 5.7** Traction power supply scheme

- for the catenary system (CS):

$$N^{rCS} = L^{CS} \times k^{S} \tag{5.6}$$

- for traction substation (TP):

$$N^{rTS} = N^{TS} \times k^{S} \tag{5.7}$$

- for electricity transmission line :

$$N^{rETL} = L^{ETL} \times k^{S} \tag{5.8}$$

where:

$L^{CS}$ ($L^{ETL}$) is the length of elements of the same design operating under the same conditions at a given polygon;

$N^{TS}$ is the number of traction substations of the same design operating under the same conditions at a given polygon;

$k^{S}$ is the coefficient taking into account the structural features of an element (type of catenary, traction substation, and electricity transmission line).

The values of conversion coefficients for each element type are given in [5].

**Table 5.3** Main characteristics of reference objects for railway power supply

| Facility (quantitative characteristic) | Reference object | Main characteristics of a reference object | Reference object elements |
|---|---|---|---|
| 1. Section of the catenary system (operational length, *N*, km) | Reference tensional length of catenary system in open lines | Tensional length of catenary system in open lines with an operational length of 1.14 km, produced according to the KS-160 projects for alternating current. | Railway catenary wires<br>Shielded conductor and line feeder<br>Pillars<br>Supporting devices<br>Fixing structures |
| | Reference tensional length of catenary system at a station | Tensional length of catenary system at station with an operational length of 1.26 km, produced according to the KS-160 projects for alternating current. | |
| 2. Traction substation (quantity, $N = 1$) | Reference traction substation | One traction substation of AC traction power supply system with a upper voltage of 110 kV | Electricity distribution devices<br>Supply transformers<br>Converter transformers (for DC power supply system) |
| 3. Section of transmission line for non-traction consumers (length, *N*, km) | Reference electricity transmission line | A section of a electricity transmission line with a length of 1 km and voltage of 10 kV with insulated wires | Wires<br>Pillar<br>Overhanging support for fastening wires |

This table implies that it is used a section of a catenary system (or electricity transmission line) of the same design and the same service life within one facility (a section of a catenary system (or electricity transmission line)). A similar reference object-element structure is used for transformer substations

### 5.2.4 Reference Object-Element Structure of Communication Facilities

According to the object-element structure of the communication complex, each real (physical) object is put in correspondence with a reference object that has the most standard properties for a given type of object (see Table 5.4).

To calculate the number of reference objects for a given facility (a group of facilities of the same type) of a communications complex at a given polygon it is required to convert existing real objects to reference objects. Moreover, each facility is expressed by an equivalent number of reference objects of a given type. The number of reference objects assigned to a real (physical) facility is not necessarily an integer.

**Table 5.4** Main characteristics of reference object-element structure for a communication complex

| Facility | Reference object | Main characteristics of the reference object |
|---|---|---|
| Facilities that make up a section of communication lines | | |
| Fiber-optic communication line | Reference FOCL | Fiber-optic line formed by a pair of optical fibers as part of 20 km long section of a fiber-optic cable. |
| Cable communication line | Reference CCL | Communication line formed by four wires with a diameter of 1.2 mm as part of 20 km long section of main feeder; and equipment for maintaining overpressure. |
| Air communication line | Reference ACL | Communication line formed by two pairs of steel uninsulated wires with a diameter of 4 mm and with length of 20 km. |
| Facilities that make up a communication center | | |
| Digital transmission system | Reference DTS | The equipment of the digital transmission system of the synchronous digital hierarchy of the STM-1 level (add-drop multiplexer) has 2 aggregate STM-1 interfaces and 63 component E12 interfaces. |
| Primary flexible multiplexing equipment | Reference PME | Digital primary flexible multiplexing equipment with 1 E12 baseband signal interface and 30 digital/analog channel termination interfaces. |
| Digital cross connect device | Reference CCD | Digital cross connect device with 64 user ports and 2 E12 trunk interfaces. |
| Structured cable system | Reference SCS | A set of telecommunication cables, cords and their switching arrangements; total length of these cables and cords is 1000 m. |

The calculation of the number of standard homogeneous or a group of sections of FOCL, CCL, ACL, DTS, PME CCD, and SCS is carried out, respectively, according to the following formulas:

For one homogeneous FOCL section or a group of homogeneous FOCL sections:

$$N^{\mathrm{rFOCL}} = \frac{\sum_{i=1}^{I} L_i Z_i}{20}; \quad N^{\mathrm{rCCL}} = \frac{\sum_{i=1}^{I} L_i Z_i k_{Di}}{20}; \quad N^{\mathrm{rACL}} = \frac{\sum_{i=1}^{I} L_i Z_i k_{Di}}{20},$$

$$N^{\mathrm{rDTS}} = \frac{\sum_{j=1}^{J} \left( \sum_{k=1}^{K_j} C_{jk} \right)}{5670}; \quad N^{\mathrm{rDTS}} = \frac{\sum_{j=1}^{J} \left( \sum_{k=1}^{K_j} C_{jk} \right)}{120}$$

(for a DTS unit or a group of DTS units of the same type, operating on CCL (ACL))

$$N^{\text{rPME}} = \frac{\sum_{j=1}^{J}\left(\sum_{k=1}^{K_j} C_{jk}\right)}{60}; \quad N^{\text{rCCD}} = \frac{\sum_{j=1}^{J}\left(\sum_{k=1}^{K_j} C_{jk}\right)}{124}; \quad N^{\text{rSCS}} = \frac{\sum_{i=1}^{I} L_i}{1000} \tag{5.9}$$

where:

$N^{\text{rFOCL}}(N^{\text{rCCL}}, N^{\text{rACL}}, N^{\text{rDTS}}, N^{\text{rPME}}, N^{\text{rCCD}}, N^{\text{rSCS}}$ ) is the number of reference sections of FOCL (CCL, ACL, DTS, PME, CCD, and SCS) corresponding to real (physical) sections

$L_i$is the length of the *i*th homogeneous section of FOCL (CCL and ACL), km

$Z_i$ is the number of pairs of optical fibers (quad of wires) used in the *i*th section of the FOCL (CCL and ACL) considered by a primary network

$I$ is the number of considered homogeneous sections of FOCL (CCL and ACL) (communication centers)

$k_{\text{D}i}$ is a conversion coefficient that takes into account the diameter of the conductors of the CCL, (ACL), used in the *i*th section of the ACL of the considered primary network

$C_{jk}$ is the equivalent information capacity of the *k*th interface of the *j*th TSE, basic digital channel

$J$ is the number of considered units of TS of one type, operating on FOCL;

$K_j$ is the total number of aggregate (linear) and component (station) interfaces of the *j*th TSE transmitting the payload data

$L$ is the total length of cables and cords of the SCS of the *i*th center, m.

## 5.3 Standardization of Locomotive Complex and Multi-unit Rolling Stock

*The basic facilities of the locomotive complex* are locomotives. At the same time, dependability and safety in the guidelines are calculated for two types of facilities: electric locomotives and diesel locomotives (steam locomotives, gas turbine locomotives, and hybrid locomotives are not considered).

*A car was adopted as the basic facility (unit) of the multi-unit rolling stock (MU).* In order to classify facilities the following types of MU are provided:

- Electric trains (DC/AC)
- Diesel trains
- Diesel-electric trains (dual-feed trains)
- Motrices.

The following units of MU are used for each type of MU:

- For electric trains and diesel-electric trains—head car, electric motor car, and trail car
- For diesel trains and railcar—head car and trail car.

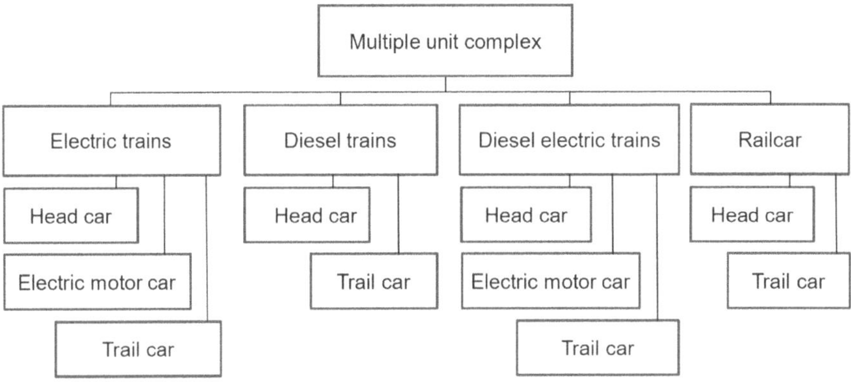


**Fig. 5.8** Composition of facilities of the multiple unit complex

The composition of facilities of the multiple unit complex is shown in Fig. 5.8.

A reference object-element structure is introduced to ensure the possibility of a comparative assessment of the indicators of MU of various series and types, with different operating life and service life and operating under different conditions.

Reference Objects Are Established for Each MU Type:

- For electric trains and diesel-electric trains—a reference head car, a reference electric motor car, and a reference trail car
- For diesel trains and railcar—a reference head car and a reference trail car.

As a reference object for a given type of MU, a car is selected from a series that has the most typical characteristics for MU of this type.

The reference object of MU is a head car (electric motor car and trail car) of a given series that passed the first 150 thousand km after its manufacturing, and operates on track sections with a straight alignment profile and a straight layout under normal climatic conditions[1]. Other head cars (electric motor car and trail car) that do not belong to the reference series, or whose operating conditions differ from those established for the reference object, are converted to the reference head car (electric motor car and trail car) using conversion coefficient.

The correspondence of the indicators of a real car of a given type of MU of a certain series under given operating conditions to the corresponding reference car is calculated using the generalized conversion coefficient $K^{\mathrm{R}}$, which is determined by several conversion coefficients according to the formula:

$$K^{\mathrm{R}} = k_1 k_2 k_3 k_4 \tag{5.10}$$

where:

$k_1$ is the coefficient of MU service life

[1]Climatic indicators of the Moscow region are taken as normal climatic conditions.

$k_2$ is the coefficient of reduction of the MU service life related to operating life
$k_3$ is the coefficient of the track profile and track layout for the section of MU operation
$k_4$ is the climatic coefficient.

The service life conversion coefficient $k_1$ and the conversion coefficient of reduction of MU service life related to operating life $k_2$ for MU of various series are estimated by empirical formulas. These formulas take into account the typical average statistical trends in the decrease in dependability related to the service life and mileage of MU:

$$k_1 = 1 + 0.2\left(\frac{t_{SL}}{t_{ASL}}\right)^2 \tag{5.11}$$

where:
$t_{SL}$ is the service life of a unit of MU, year
$t_{ASL}$ is the assigned service life of MU unit of the given series, year.

$$k_2 = 1 + 0.16 * 10^{-\left(\frac{50x}{x_{max}}\right)^2} + 0.2\left(\frac{x}{x_{max}}\right)^4 \tag{5.12}$$

where:
$x$ is the operating life of a unit of MU, thousand km mileage
$x_{max}$ is the maximum operating life of a unit of MU corresponding to the end of the assigned service life.

The conversion coefficient for the track profile and layout for the section of MU operation ($k_3$) is calculated as follows. The considered section of the MU operation is divided into elementary homogeneous sections, within each of which the track has approximately the same parameters (track gradient and swivel radius). The coefficient for the track profile and layout ($k_3$) is calculated as follows:

$$k_3 = 1 + \frac{\sum_{j=1}^{J}\left(\frac{z_j}{1000} + w_j\right)L_j}{\sum_{j=1}^{J} L_j} \tag{5.13}$$

where:
$z_j$ is the track gradient of the $j$th elementary track section, ‰
$w_j = \frac{700}{R_j}$ is the coefficient of resistance to movement on the $j$th elementary curved alignment section of the track, where $R_j$ is the swivel radius, m (for a straight elementary section, we assume $w_j = 0$)
$L_j$ is the length of the $j$th elementary track section, km

$J$ is the number of elementary track sections on the considered section of the MU operation.

The climatic coefficient is the ratio of the daily maximum amount of rainfall in a given climatic region to the daily maximum amount of rainfall in that climatic region, which is taken as the initial one (i.e., the climatic coefficient of which is conventionally assumed to be equal to 1). In Russia, the region of the middle zone of the European part (Central region) is conventionally taken as the initial climatic one. *The climatic coefficient* $k_4$ is calculated as the product of several constituent coefficients:

$$k_4 = k_{41}k_{42}k_{43}k_{44} \tag{5.14}$$

where:

$k_{41}$ is the coefficient that takes into account the proportion of time during the year with air temperatures less than minus 20 °C

$k_{42}$ is the coefficient taking into account the proportion of time during the year with air temperatures above +30 °C

$k_{43}$ is the coefficient taking into account the proportion of time during the year with air humidity over 85%

$k_{44}$ is the coefficient that takes into account the annual precipitation level.

The climatic indicators of the Moscow region are taken as normal climatic conditions. Coefficients $k_{41}$ and $k_{42}$ are determined by the formula:

$$k_{41} = 1 + \alpha_{41}\frac{n_{41} - 7.5}{365}, \qquad k_{42} = 1 + \alpha_{42}\frac{n_{42}}{365} \tag{5.15}$$

where:

$n_{41}$ is the number of days in a year with an average air temperature of less than minus 20 °C in the areas of MU circulation

$\alpha_{41}$ is the proportion of the power loss of the MU at an air temperature of less than minus 20 °C

$n_{42}$ is the number of days per year with an average air temperature of more than plus 30 °C in the areas of MU circulation

$\alpha_{42}$ is the proportion of the power loss of the MU at an air temperature of more than plus 30 °C.

Coefficients $\alpha_{41}$ and $\alpha_{42}$ are set on the basis of expert estimates and can be corrected if necessary. It was preliminarily established that $\alpha_{41} = 0.06$ and $\alpha_{42} = 0.09$.

Coefficient $k_{43}$ is determined by the formula:

$$k_{43} = 1 + \alpha_{43}\frac{n_{43} - 67.2}{365} \tag{5.16}$$

where:

$n_{43}$ is the number of days per year with an air humidity of more than 85% in the areas of MU circulation

$\alpha_{43}$ is the proportion of the decrease in the quality of MU operation under conditions of air humidity over 85%.

Coefficient $\alpha_{43}$ is set on the basis of expert estimates and can be corrected if necessary. It was found that $\alpha_{43} = 0.26$.

The coefficient $k_{44}$ is determined by the formula:

$$k_{44} = 1 + \alpha_{44}(n_{44} - 582) \tag{5.17}$$

where:

$n_{44}$ is the average annual amount of precipitation, mm, in the areas of MU circulation

$\alpha_{44}$ is the share of the decrease in the quality of the MU functioning under conditions of increased precipitation The coefficient $\alpha_{44}$ was determined on the basis of expert estimates and can be corrected if necessary. It was found that $\alpha_{44} = 0.0001$.

## 5.4 Fundamentals of Normalization of Dependability Indicators

### *5.4.1 The Goal of Normalization of Dependability Indicators*

Dependability normalization is the specification (in technical or other documentation) of quantitative or qualitative requirements for dependability. Therefore, normalization sets acceptable limits for changes of a controlled characteristic. A dependability indicator is a characteristic (as a rule, quantitative) of one or several properties comprising the dependability of a technical system (facility). The values of dependability indicators can be normative or factual. They can be determined by calculation methods, on the basis of test or experiment data, operations data or by extrapolation. Factual values of dependability indicators during the process of operation of a technical system are obtained based on the analysis of statistical data on a system's failures and time to its recovery. As far as normative values of dependability indicators, they are as a rule specified in a quantitative way at the design stage of a facility. For most facilities one applies a normalization probabilistic approach when one normalizes and ensures a required economically justified level of probabilistic dependability indicators that is afterwards controlled by dependability tests and kept by a maintenance system. The exclusion is safety critical facilities with catastrophic failure consequences, whose failures are not acceptable (this book does not consider such facilities since they belong to the field of functional safety).

The results of evaluation of a technical facility's factual state allow making a decision [6] on its further operation (operation continuation, maintenance assignment, decommissioning, a facility's replacement, etc.). Under the conditions of

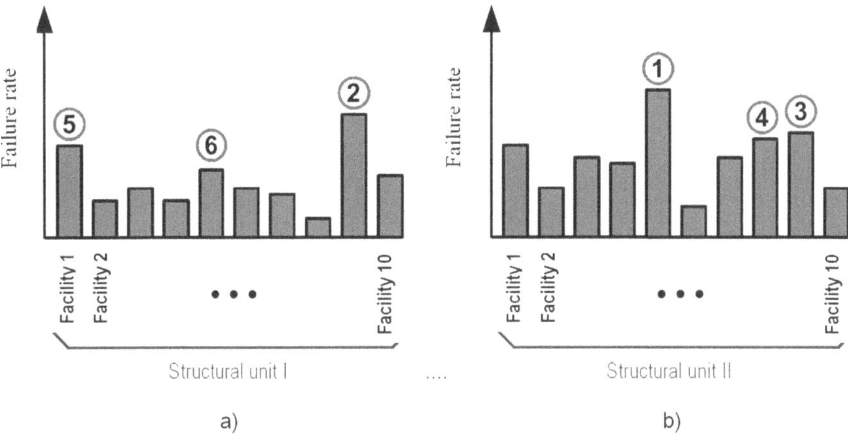


**Fig. 5.9** Example of determination of facilities' order of priority for repair assignment (without normalization)

resource limits, it is vital to identify most "problematic" facilities that require primary investments.

Figure 5.9a, b shows an example of determination of priority levels of railway infrastructure facilities requiring the enhancement of dependability—for example, by assignment and execution of repair—for two enterprise units, where facilities of one type are under different operating conditions. In this example, we assume that there are funds reserved for repair of 6 facilities in these two enterprise units.

Figure 5.9a shows that based on factual values of a dependability indicator (for example, a failure rate), that reflects the current state of the facility in operation, we can identify those facilities that require repair assignment as a priority with the size of an allocated investment taken into account. In this case, if normative values are not available, facilities are chosen by the criterion of the worst indicator value. When introducing normalization of indicators one should take into account non-similar operating conditions for facilities in different enterprise units, which are determined by differences in climatic factors, technical capabilities for maintenance and repair, staffing levels, grades of tear and wear of facilities, and requirements for their productivity (for example, with different sizes of train traffic). In this case facilities will be chosen for repair assignment by the criterion of an indicator's deviation to the worse side from a normative value (Fig. 5.10a, b).

Obviously, introduction of normative indicators considering operation conditions and other factors of enterprise units' activities improves targeted investment allocation for maintenance of facilities, which allows fulfilling the requirement of uninterruptible transportation under the conditions of resource scarcity [6].

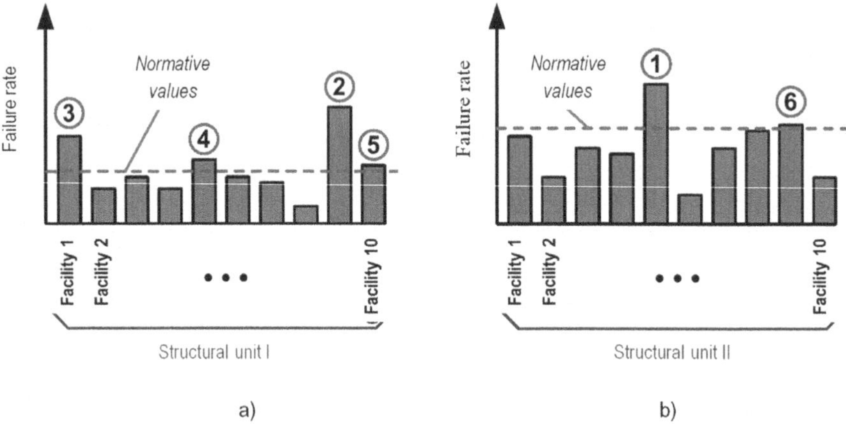


**Fig. 5.10** Example of determination of facilities order of priority for repair assignment (with normalization)

## 5.4.2 *Interests of the Service User and Supplier*

In case when a technical system is involved in providing services (for example, a railway infrastructure facility ensures transportation process execution), normative values of dependability indicators shall consider relations between a supplier and a user of a service (for example, an enterprise unit in charge of the functioning of a railway infrastructure facility and an enterprise unit executing transportation process).

It is worth to note that this scheme presents an inevitable conflict between the interests of a user and a supplier of a service. From the one hand, a user is interested that there would not be any failures of a facility providing a service at all; this would allow him to execute his activities with no risk related to a facility failure (for example, a risk of train-hours loss due to the failure of a railway infrastructure facility). From the other hand, a supplier is interested in reducing the costs of a service, thus increasing the operating profit, but a reduction of costs inevitably causes increased failure rates. Normalization of a facility's dependability indicators shall in essence ensure a compromise between the interests of a supplier who seeks to provide a service under the conditions of resource limits and the interests of a user who seeks to have a service of high quality with the lowest expenditures.

The situation in question is similar to the situation when a user receives a product batch from a supplier and where the unambiguity of mutual acknowledgment of a product's quality by a supplier and a user is in most cases regulated by methods of statistical acceptance tests. And the relations between a supplier and a user characterize an acceptable level of quality $x_\alpha$ (the maximum acceptable value of defective items share in a batch) and an unacceptable quality level $x_\beta$ (the boundary of defective items share for attributing a batch as defective), where $x_\alpha \leq x_\beta$ (Fig. 5.11). Therefore, the area of a user's interests is $x \leq x_\beta$ and the area of a

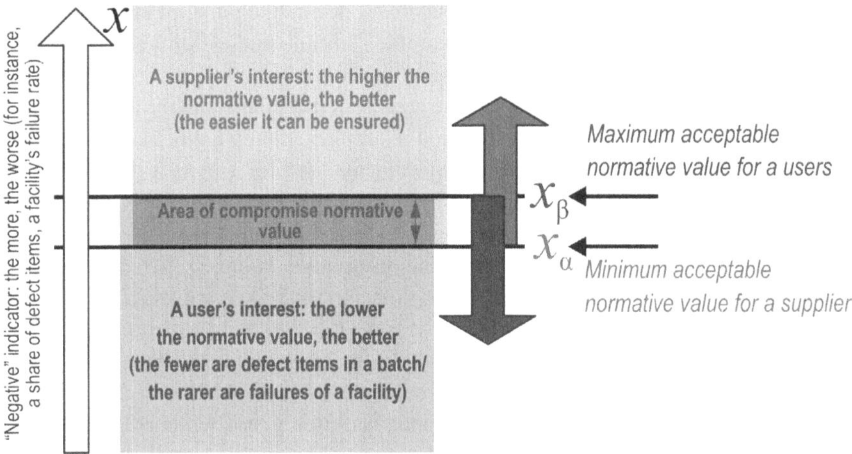


**Fig. 5.11** Areas of a service user's and a service supplier's interests

supplier's interests is $x > x_\alpha$; it is obvious that the two areas cross each other that being a prerequisite condition for the existence of compromise between both interests. The area of an attribute value $x$ under $x_\alpha$ is "acceptance region," that above $x_\beta$ is "unacceptance region," and that between $x_\alpha$ and $x_\beta$ is "uncertainty region."

Note that the application of two points ($x_\alpha$, $x_\beta$) as threshold values is a general practice for a sequential selective test [7], where a conclusion on usability (or non-usability) of a batch is made on the basis of a defective items share in a selection that is a part of a batch volume. For this process, acceptable and unacceptable levels are set using confident intervals. A probability that a share of defective items in the whole batch is not larger than $x_\alpha$, when an upper confident interval of an unacceptable level is exceeded for a selection test, is a supplier's risk; vice versa, a probability that a share of defective items in the whole batch is larger than $x_\beta$, when a lower confident interval of an acceptable level is reached for a selection test, is a user's risk. From Fig. 5.11, it follows that a service supplier can guarantee that the factual value $x$ of an indicator will be above the threshold value $x_\alpha$ with a high degree of confidence (probability), for example, $P_\alpha = P\{x > x_\alpha\} > 0.95$ (supplier's risk $Q_\alpha = 1 - P_\alpha \leq 0.05$) (GOST R ISO 8422-2011. Statistical methods. Sequential plans of selective tests as per alternative attribute); a service user expects that the factual value x of an indicator will not be higher than the threshold value $x_\beta$ with a high degree of confidence (probability), for example, $P_\beta = P\{x \leq x_\beta\} > 0.9$ (user's risk $Q_\beta = 1 - P_\beta \leq 0.1$).

Under the real conditions of a technical system's operation, the task of evaluating its compliance with specified requirements of dependability is often brought down to comparison of the value of a factual dependability indicator obtained for some period of observance of statistical operational data with a normative value specified in technical or other documentation. In this case the presence of "uncertainty region" will complicate estimation making it ambiguous. That is why technical documentation for a facility generally contains a normative value of an indicator in form of a single threshold value (for example, "mean time to failure shall be not lower than 30,000 h, maintenance inclusive").

Let a single threshold normative value $x_\eta$ is specified for a facility dependability indicator by agreement between a user and a supplier, then we will assume that for $x \leq x_\eta$ this facility complies with the requirements, and for $x > x_\eta$ it does not. It is obvious (see Fig. 5.11 ) that when transiting from two threshold levels to one it is reasonable to comply with the condition $x_\alpha < x_\eta \leq x_\beta$ ($x_\eta$ belongs to the area of "compromise values"), in which case the normative value $x_\eta$ for the attribute $x$ satisfies to the requirements of both a service user and a service supplier.

In the case of a single threshold value, the risk $Q_\eta = P\{x > x_\eta\}$ of non-compliance of an indicator with specified requirements is in fact split between a user and a supplier of a service according to their agreement (for example, the exceedance of a normative value at one interval of observance is a user's risk, while at two or more consecutive intervals of observance it is the responsibility of a supplier).

One of the ways of normalizing dependability indicators used in the global practice (in particular, in the power supply field) is the normalization based on past experience (analysis of factual data on dependability) [8]. Given the availability of such data on railway transport, we will consider a further task as a choice and justification of the value $x_\eta$ using existing statistical data on the operation of a facility during some interval of observance, assuming that in general these indicators of a facility's dependability may be evaluated for this interval of observance as acceptable for a service user.

### 5.4.3 *Analysis of Statistical Data and Evaluation of Their Sufficiency*

As it was noted earlier, the factual values of dependability indicators are random values. For example, for a facility's failure rate (number of failures per time unit) the statistics presents a time series of discrete values—for instance, this is a sequence of failure rate values per each annual interval of observance for several years.

A random value is fully defined by a distribution law, for discrete values this is a distribution series or a discrete distribution function. A distribution series (or a discrete distribution function) presents a table of possible values of a random size with respective probabilities.

There are a great number of various theoretical laws of distribution (uniform, Bernoulli, Cauchy, Poisson, normal, lognormal, Gumbel, Jonson, 13 Pearson's curved distributions, etc.) [9]. However, in practice one often deals with statistical material of rather a limited volume, and it is not always possible to identify a concrete distribution law for a random value based on this volume. In such cases it is necessary to describe the behavior of a random value by numeric characteristics.

**Table 5.5** Initial time series of a facility's failure rate

| Observation year | 2008 | 2009 | 2010 | 2011 | 2012 | 2013 | 2014 | 2015 | 2016 |
|---|---|---|---|---|---|---|---|---|---|
| Failure rate, $x_i$, 1/year | 34 | 37 | 24 | 17 | 12 | 9 | 13 | 43 | 36 |

For engineering calculations and scientific researches one uses empirical curved distributions of random values characteristics. When constructing such curves, major stages are ranking of an initial time series and estimation of its empirical sufficiency. Solving the first of these tasks presents no difficulties, whereas for the second it is necessary to take into account that some formulas for estimation of sufficiency lead to systematic errors and give different values of random errors.

Scarcity function $q(x)$ is an analog of distribution function $F(x)$ and characterizes a probability that the value of an argument exceeds a specified threshold value [10] based on theoretical researches and results of testing defined a formula, which gives efficient, nonbiased, and effective values of scarcity estimates of the $i$th ($i = 1...n$) member of a discrete sample ranked in descending order (i.e., of probabilities $q_i$ that the factual value $x$ exceeds the value of $x_i$ series member):

$$q_i = P\{x > x_i\} = \frac{i}{n+1} \tag{5.18}$$

where $n$ is a number of series members.

Let us consider the algorithm comprising the ranking of an initial time series, the estimation of its empirical sufficiency, and the approximation by a theoretical distribution law using the statistical data on failures of primary railway telecommunications network facilities for the years of 2008–2016 (Table 5.5, the data submitted by the Central telecommunications station—JSC RZD branch).

1. An initial series is ranked in order of descending of an indicator's values. Instead of observance years we introduce conditional numbers of a ranked series' members (1, 2, 3, ...)
2. For each member of a ranked series we calculate values $q_i$ of scarcity function using formula (5.18)
3. Then we calculate mathematical expectation of series members
4. For each member of a ranked series we calculate a modulus coefficient equal to a relation of a series member's value to a series' mathematical expectation.

   As a result, we have Table 5.6.
5. For refinement of values of distribution quantiles ($q$), especially at levels lower than 0.2, that are of practical interest, we make approximation of a series of modulus coefficients (Table 5.6) using one of the theoretical distribution laws. As an example, let us consider approximation by a three-parameter gamma distribution [11] that has been in particular applied in hydrological calculations [12], and calculations of structures' service life under random load flows.

**Table 5.6** Ranked time series with scarcity estimates and modulus coefficients

| Item No., $i$ | 1 | 2 | 3 | 4 | 5 | 6 | 7 | 8 | 9 |
|---|---|---|---|---|---|---|---|---|---|
| Failure rate, $x_i$, 1/year | 83 | 76 | 37 | 34 | 24 | 14 | 13 | 12 | 9 |
| Mod. coefficient, $k_i$ | 2.4735 | 2.2649 | 1.1026 | 1.0132 | 0.7152 | 0.4172 | 0.3874 | 0.3576 | 0.2682 |
| Scarcity, $q_i$ | 0.1 | 0.2 | 0.3 | 0.4 | 0.5 | 0.6 | 0.7 | 0.8 | 0.9 |

Using the modular coefficients $k_i$ from Table 5.6, the coefficient $C_\nu$ of variation of the series and the ratio of $C_{s\nu}$ of the coefficient of asymmetry of the series to the coefficient of variation of the series are calculated:

$$C_\nu = \sqrt{\frac{1}{n-1}\sum_{i=1}^{n}\left(\frac{x_i}{x}-1\right)^2};\, C_{s\nu} = \frac{n\sum_{i=1}^{L}\left(\frac{x_i}{x}-1\right)^3}{(n-1)(n-2)C_\nu^4} \quad (5.19)$$

where if in (2) we obtain the value $C_\nu < 0.1$, then before calculating $C_{s\nu}$, as well as for further usage, we assume that $C_\nu = 0.1$ (as for series with a very small variation it is complicated to define distribution quantiles). After calculation the value $C_\nu$ is approximated to multiplicity 0.1 (0.1; 0.2; 0.3 ...), while the value $C_{s\nu}$ is approximated to multiplicity 0.5 (0; ±0.5; ±1.0; ±1.5; ...) to allow the application of existing table distribution function values since their analytical calculation is very complicated.

Using Table values of three-parameter gamma distribution functions [13] for a specified scarcity probability $q_i$, we define the ordinate $m_i$ in form of a modulus coefficient (the mentioned tables contain values of a distribution functions for various values $C_\nu$ and most widely-spread relations $C_s/C_\nu$).

In the example in question for values $C_\nu = 0.5$ and $C_{s\nu} = 0$ obtained using formulas (5.19), we have a series of values of function ordinates as modulus coefficients $m_i(p_i)$, including additional values at the boundaries of a function (Table 5.7). In order to obtain quantitative values $y_i$ of failure rate that will be exceeded with the probability $q_i$, we should multiply modulus coefficients $m_i$ by the value of mathematical expectation of a ranked series from Table 5.6 (the results are summarized in Table 5.7).

The estimation of approximation reliability was made using a coefficient of an empirical linear correlation $x_i(q_i)$ and a function chosen as per this method $y_i(q_i)$ (for $i = 1 \ldots 9$). We obtained the value of a linear correlation coefficient as 0.974, which is close to 1, thus confirming the closeness of the chosen function to the initial series with a high reliability.

Figure 5.12 presents a graph of an empirical series (points) and an approximating function of a three-parameter gamma distribution (solid line).

### 5.4.4 Choice and Justification of Normative Indicator Value

The results of sufficiency estimation obtained above (see Table 5.7) can be applied for defining a threshold value $x_\eta$ for a specified level of risk $Q_\eta$ agreed between a supplier and a user of a service or vice versa for estimating risk $Q_\eta$ based on a specified value $x_\eta$.

Let us consider a case when for a specified risk level of non-compliance with a normative value (for example, $Q_\eta = 0.1$) we have to define a normative value $x_\eta$ of a

**Table 5.7** Example of an approximated time series with scarcity estimates and modulus coefficients ($C_{\nu} = 0.8$ и $C_{s\nu} = 1.4$)

| Item No., $i$ | – | – | 1 | 2 | 3 | 4 | 5 | 6 | 7 | 8 | 9 | – | – |
|---|---|---|---|---|---|---|---|---|---|---|---|---|---|
| $q_i$ | 0.01 | 0.05 | 0.1 | 0.2 | 0.3 | 0.4 | 0.5 | 0.6 | 0.7 | 0.8 | 0.9 | 0.95 | 0.99 |
| Mod. coefficient, $m_i$ | 2.01 | 1.8 | 1.66 | 1.47 | 1.31 | 1.16 | 1.01 | 0.86 | 0.69 | 0.51 | 0.31 | 0.18 | 0.06 |
| Failure rate, $y_i$, 1/year | 50.25 | 45 | 41.5 | 36.75 | 32.75 | 29 | 25.25 | 21.38 | 17.25 | 12.78 | 7.63 | 4.55 | 1.38 |

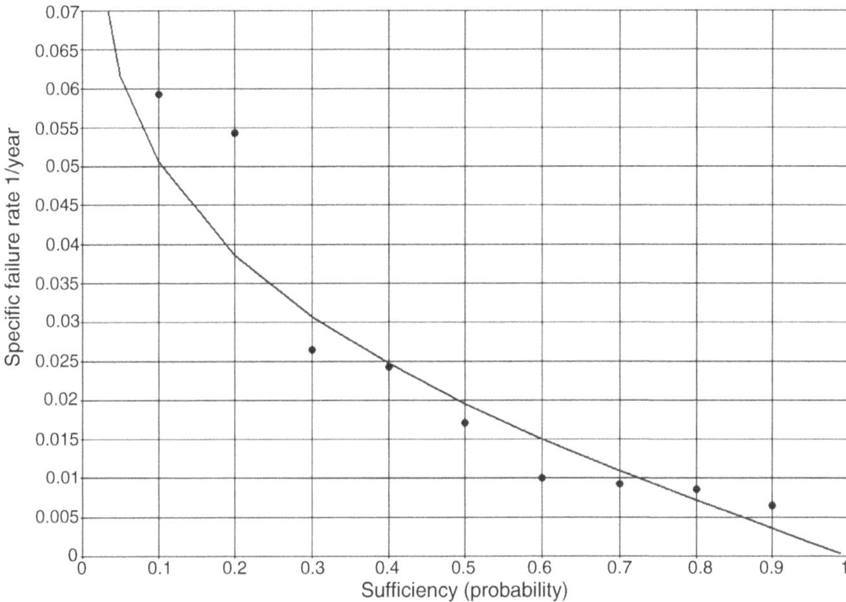


**Fig. 5.12** Graph of the empirical series (points) and the approximation function (solid line)

dependability indicator (in our case it is a facility's failure rate). Let us estimate the quantile of a sufficiency function that corresponds to a specified risk ($q_i = Q_\eta = 0.1$). According to the data of Table 5.7 we have

$$y_i(Q_\eta) = y(q_1 = 0,1) = 70,8 \approx 71.$$

Therefore, as an indicator's normative value we can take a failure rate equal to 71 1/year, which will be not ensured with a risk of 0.1. If by agreement between a supplier and a user of a service there is a specified normative value of dependability indicator, in a similar way based on the obtained results of sufficiency estimation (see Table 5.7), one can define risk of non-compliance of an indicator with specified requirements.

In any case an agreement between a supplier and a user of a service shall foresee both the specification of a normative value of all dependability indicators in question and the specification of risk levels for nonfulfillment of these normative values as well as the procedure of splitting of responsibility between a supplier and a user of a service.

The method allows defining a relation between a value of a dependability normative indicator and a risk of its nonfulfillment by objective criteria based on factual capabilities of operated facilities that are estimated as per existing statistical data for the past periods.

To choose and justify the normalized value of a dependability indicator, the authors have studied the relations between a service supplier and a service user, have analyzed statistics using the method of estimation of empirical sufficiency of a raw data series as well as approximation of an ordered initial series by a three-parameter gamma distribution.

The proposed approach allows establishing a correlation between a normalized value and a risk of its violation via a function of sufficiency, which can be obtained on the basis of existing statistical data on a facility's dependability for the past periods. This correlation makes it possible to guarantee the ensuring of compliance of factual and normalized indicator values with a specified risk level for a facility working in normal mode.

### *5.4.5 Normalization of Key Dependability Indicators*

Above, we considered the normalization of the dependability indicator using the example of the specific failure rate (negative indicator). The convenience of using a negative indicator is due to the following:

- The point rating scale, according to which the performance of structural units is assessed, is also negative
- The sufficiency function used in this methodology is focused on the indicator, the lower limit of which is equal to 0, and the upper limit is not limited; this property is well suited to negative indicators such as failure rate and mean time to recovery.

Next, let us consider the method of normalization of the remaining key indicators in respect to the track superstructure.

The list of key indicators of the track superstructure dependability is presented in Table 5.8.

**Table 5.8** List of key indicators of track superstructure dependability

| Indicator name | Legend | Type of indicator | Based on following indicator | Related negative indicator |
|---|---|---|---|---|
| Specific mean time to failure | $\widehat{X}_0, \widehat{X}_0^t$ | Positive | • (Primary) | Specific failure rate $\widehat{\lambda}, \widehat{\lambda}^t$ |
| Specific failure rate | $\widehat{\lambda}, \widehat{\lambda}^t$ | Negative | • (Primary) | – |
| Specific probability of failure-free operation | $P(x)$ | Positive | Specific failure rate $\widehat{\lambda}$ | – |
| Mean time to recovery | $T_R$ | Negative | • (Primary) | – |
| Specific availability coefficient | $\widehat{K}_A$ | Positive | Specific mean time to failure $\widehat{X}_0^t$, mean time to recovery $T_R$ | – |
| Specific downtime coefficient | $\widehat{K}_D$ | Negative | • | – |

1. *Specific mean time to failure.* Since this indicator in the system of key indicators has a related "negative" indicator—the specific rate of failures—then its normalization is carried out in the following way. The statistical series of mean time to failure values is transformed into a series of values of related "negative" indicator (failure rate); each $i$th ($i = 1 \ldots n$) member of the series is transformed:

$$\widehat{\lambda}_i = \frac{1}{\widehat{X}_{0i}} \qquad \widehat{\lambda}^t_i = 1/\widehat{X}^t_{0i} \tag{5.20}$$

Then we normalize of the value of the specific failure rate, as in the above example. After that, the obtained lower and upper recommended limits for the specific failure rate are converted to mean time to failure values, respectively:

$$\widehat{X}_{0\text{н}.h} = \frac{1}{\widehat{\lambda}_{\text{н}.l}};\widehat{X}_{0\text{н}.l} = \frac{1}{\widehat{\lambda}_{\text{н}.h}};\widehat{X}^t_{0\text{н}.h} = \frac{1}{\widehat{\lambda}^t_{\text{н}.l}};\widehat{X}^t_{0\text{н}.l} = 1/\widehat{\lambda}^t_{\text{н}.h} \tag{5.21}$$

2. *Specific probability of failure-free operation.* This indicator (under the condition of the operating time interval $\Delta x$, or time $\Delta t$ is chosen) directly depends on the specific failure rate. Its normalization is carried out on the basis of the already normalized lower and upper limits of the specific failure rate:

$$\widehat{P}_{\text{н}.h}(x) = -\exp\left(\widehat{\lambda}_{\text{н}.l}\Delta x\right),\ \widehat{P}_{\text{н}.l}(x) = -\exp\left(\widehat{\lambda}_{\text{н}.h}\Delta x\right) \tag{5.22}$$

The calculation of norms for the specific probability of no-failure operation for the operating time interval in the form of time ($\Delta t$) is performed in the same way, using the parameters $\widehat{\lambda}^t_{\text{н}.l}$ and $\widehat{\lambda}^t_{\text{н}.h}$.

3. *Mean time to recovery.* The normalization of this indicator is carried out in the same way as for the failure rate in the example considered above.
4. *Specific availability coefficient.* Since this indicator is based on two primary indicators, its normalization is carried out according to pre-established norms for the specific mean time to failure and mean time to recovery:

$$\widehat{K}_{\text{гн}.h} = \frac{X^t_{0\text{н}.h}}{X^t_{0\text{н}.h} + T_{\text{вн}.l}} \quad ; \qquad \widehat{K}_{\text{гн}.l} = \frac{X^t_{0\text{н}.l}}{X^t_{0\text{н}.l} + T_{\text{вн}.h}} \tag{5.23}$$

5. *The downtime coefficient* is normalized in the same way as the failure rate in the above example.

## References

1. Metodika rascheta pokazatelej nadezhnosti i bezopasnosti funkcionirovaniya etalonnyh ob"ektov putevogo hozyajstva OAO "RZD" (Method for calculating dependability and safety indicators of operation of the reference objects of JSC "Russian Railways" track facilities), Moscow, RZD (2011)
2. Metodika ocenki pokazatelej deyatel'nosti podrazdelenij proizvodstvennogo bloka OAO "RZD" po sostoyaniyu infrastruktury i effektivnosti raboty zheleznodorozhnyh uchastkov (Method for assessing the performance indicators of the divisions of the production unit of JSC "Russian Railways" in terms of the state of the infrastructure and the efficiency of the railway sections functioning), Moscow, RZD (2014)
3. Metodika ocenki deyatel'nosti strukturnyh podrazdelenij remontnogo kompleksa putevogo hozyajstva OAO "RZD" po pokazatelyam nadezhnosti i bezopasnosti funkcionirovaniya zheleznodorozhnogo puti i putevyh mashin (Method for assessing the activity of structural divisions of the repair complex of track facilities of JSC "Russian Railways" in terms of dependability and safety indicators of the railway track and track machines functioning), Moscow, RZD (2012)
4. Metodika rascheta pokazatelej ekspluatacionnoj nadezhnosti ob'ektov hozyajstva avtomatiki i telemekhaniki (Method for calculating indicators of operational dependability of signalling and remote control facilities), Moscow, RZD (2011)
5. Metodika rascheta pokazatelej ekspluatacionnoj nadezhnosti, intensivnosti otkazov, narabotki na otkaz i koefficienta gotovnosti etalonnyh ob'ektov hozyajstva elektrifikacii i elektrosnabzheniya OAO "RZD" (Method for calculating indicators of operational dependability, failure rate, MTBF and availability coefficient of reference objects of the electrification and power supply economy of JSC "Russian Railways"), Moscow, RZD (2011)
6. Koncepciya kompleksnogo upravleniya nadezhnost'yu, riskami, stoimost'yu zhiznennogo cikla na zheleznodorozhnom transporte (The concept of integrated management of reliability, risks, life cycle cost in rail transport), Moscow, RZD (2010)
7. Shubinsky, I.B., Novozhilov, E.O.: Method of normalization of dependability indicators of railway transport facilities. Dependability Journal. **19**(4), 17–23 (2019)
8. Rudenko, Y.N.: O podhodah k normirovaniyu pokazatelej nadezhnosti elektrosnabzheniya potrebitelej (Approaches to the normalization of the dependability indicators of electric power supply to consumers). Izvestiya Akademii nauk SSSR. Energetika i transport. **1**, 14–23 (1975)
9. Litvinenko, R.S., Pavlov, P.P., Idiyatullin, R.G.: Practical application of continuous distribution laws in the theory of dependability of technical systems. Dependability Journal. **16**(4), 17–23 (2016)
10. David, H.: Poryadkovye statistiki (Order statistics). Nauka, Moscow (1979)
11. Vadzinsky, R.N.: Spravochnik po veroyatnostnym raspredeleniyam (Handbook of probability distribution). Nauka, Saint-Petersburg (2001)
12. Sikan, A.V.: Metody statisticheskoj obrabotki gidrometeorologicheskoj informacii (Methods of statistical processing of hydrometeorological information). RGGMU, Saint Petersburg (2007)

# Chapter 6
# Fundamentals of Management of Technical and Industrial Risks on Railway Transport

## 6.1 Definitions, Classification of Risks, and Safety Principles

### *6.1.1 Risk Definitions*

In the opinion of the outstanding etymologist and specialist in Slavic studies M. Fasmer [1] the word "risk" is a borrowing from the French or Italian languages (French: risque, Italian: risico). This word traces back to the ancient Greek ῥιζικόν meaning "cliff" and ῥίζα meaning "root." Also "take risks" originates from French risquer or Italian risicare which originally means "maneuver between the rocks."

Risk is an activity associated with overcoming uncertainty in a situation of inevitable choice, and in the process of this activity, it is possible to assess quantitatively and qualitatively the probability of achieving the intended result, failure, and digressing from a goal. In general, risk is understood to be a possibility of some adverse event to occur which results in various types of losses (for example, physical injury, loss of property, income below the expected level, etc.). As for business activity, by a "risk" is commonly meant to be a probability of a partial loss of resources by an enterprise, a loss of income, or an incurrence of additional costs as a result of certain production and financial activities.

The definition of a risk is distinctive depending on the field of its application. These include the following definitions. *Risk is a characteristic of a situation where an outcome is uncertain, provided always that there are adverse consequences* [2].

When defining a "risk" in *a restricted sense it means a quantitative assessment of hazards that is determined by the frequency of one event occurrence in case of occurrence of another event.*

Risk *is an indefinite event or condition that, if it happens, has a positive or negative effect on a company's reputation and results in profit or financial losses.*

 
I. B. Shubinsky, A. M. Zamyshlaev, *Technical Asset Management for Railway Transport*, International Series in Operations Research & Management Science 322,
https://doi.org/10.1007/978-3-030-90029-8_6

Risk *is a probability of possible undesired loss of something in case of unfortunate combination of circumstances.*

Risk *is a probability of losing control over dangerous factor and the severity of the consequences, expressed by the level of its intensity.*

Risk *is the probability multiplied by loss.* The names of the events leading to damage are a list of risk factors. The frequency of event occurrence is the basis for determining the probability of risk.

The following terminological definitions of risk were standardized:

**Risk** ***is a combination of the probability and severity of consequences (harm)*** (IEC 62278)

**Risk** ***is a combination of the probability of event occurrence and its consequences*** (GOST 33433, IEC 60300-3-9:1995)

**Risk** ***is an effect of uncertainty on objectives*** (ISO 31000)

**Risk** ***is a consequence of the influence of uncertainty on the achievement of set goals*** (ISO Guide 73:2009).

A "risk" has specific properties as follows from the definitions.

*Uncertainty*. Risk exists if, and only if, not the only scenario is possible.

*Damage*. The risk exists when the outcome can lead to damage (loss) or other negative (only negative!) consequence.

*Availability of analysis*. The risk exists only when a subjective opinion of the "person making a suggestion" about the situation is formed and a qualitative or quantitative evaluation of the negative event of the future period is given (otherwise it is a threat or danger).

*Significance*. Risk exists when an expected event is of practical importance and affects the interests of at least one subject. There is no risk without a risk subject.

As follows from the standards IEC 62278, GOST 33433, IEC 60300-3-9:1995, the quantitative determination of the risk level $R$ involves its expression using two values—the frequency or probability $p$ of an undesired event and the size of its consequences $C$. The ISO Guide 73:2009 implies a similar approach to determining the level of risk. In practice, typically when determining the level of risk, the frequency (probability) and the size of the consequences act as multipliers $R = F * C$ (or $R = p * C$ if one predicts the occurrence of an undesirable event with probability $p$). Lines formed by a set of points with a given constant risk level $R$ are hyperbolas $=R/C$. If we denote a constant level of risk as *const*, then these hyperbolic curves are denoted as $F = \text{const}/C$. In Fig. 6.1, the ordinate shows the frequency of undesired events, and the abscissa shows possible damage. It is presented a hyperbolic curve that illustrates a constant level of risk for various combinations of risk components $R$ and $C$. Fig 6.1 illustrates that the risk graph is equivalently determined by the values $F$ and $C$. In case of unlikely (rare in frequency) undesirable events but with great damage, the risk can be at the level of the situation when undesirable events occur frequently, but the damage from each event can be relatively small.

In case of one knows consequences (for example, injury or fatalities), the risk can only be expressed by probability or frequency. It is also possible to use the definition of risk as the probability of exceeding a certain limit by some random parameter (threshold).

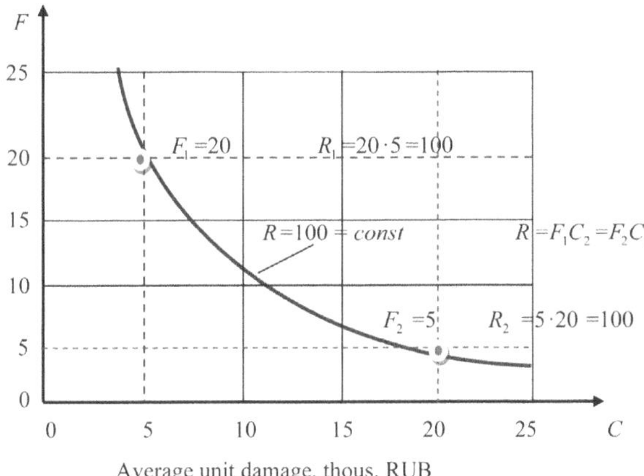


**Fig. 6.1** Graph of dependence of risk on the frequency of an undesired event and the amount of associated damage

## 6.1.2 Risk Classification

Each undesired event can occur in relation to a certain object of risk. All types of railway manufacturing company's risk can be conditionally divided into three groups:

- Risks related directly to the production activities of a company (transportation process): operational, technosphere, occupational risks
- Risks associated with the economic activities of a company (provision of freight and passenger transportation services, communication services, power supply service for non-production enterprises, employees support service, passenger service, cargo handling service, service of transport; insurance services, etc.): financial, commercial, insurance, market, and legal risks
- Risks associated with the company's image: political, social, and legal risks.

Figure 6.2 shows the classification of the main risks.

The URRAN system assesses the first group of risks that are directly related to the transportation process. Among these, first of all, are operational risks comprising technical and technological risks. *Technical risks* are associated with undesired events due to failures of railway facilities. *Technological risks* are caused by technological violations (violations of instructions, regulations, and rules for the implementation of the transportation process). The relevant regulations of the Customs Union (see Fig. 6.2) give the safety requirements for the operation of rolling stock, infrastructure facilities, and high-speed traffic facilities. When ensuring the safety of the railway transportation process, the so-called technosphere risks should be taken into account. These include environmental, industrial, and fire risks (Fig. 6.2). Industrial risks are due to the operation of hazardous production facilities. *Fire risks* give cause for particular concern when implementing transportation process since there are a large number of ignition factors related to locomotives, signal boxes, computer centers, etc.

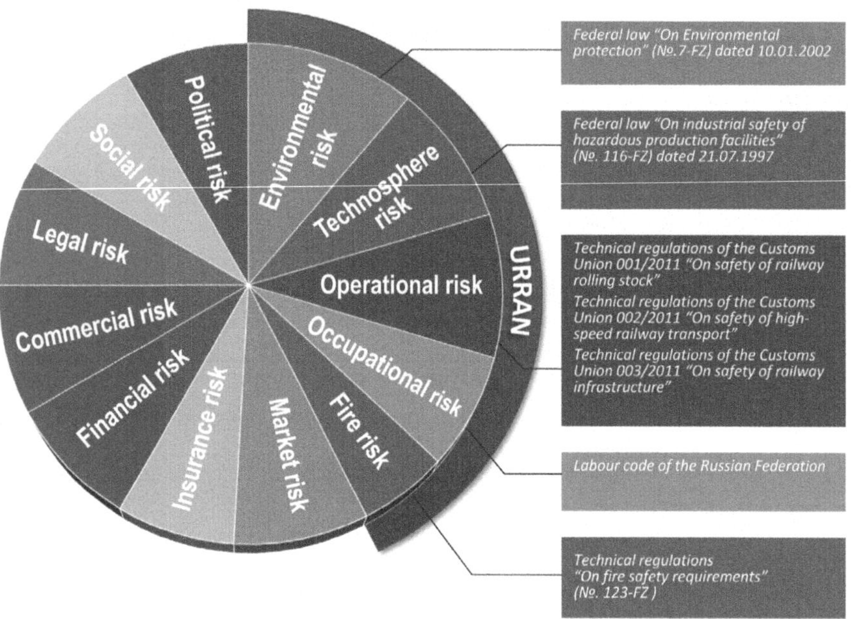


**Fig. 6.2** Classification of risks on railway transport

Occupational risks have a great influence on the safety of transportation process. These risks are associated with injuries and even employee fatalities on the railways due to insufficient professional training of employees, poor labor management, violations of rules and instructions when performing maintenance and repair of equipment, as well as transporting maintenance personnel and equipment. Both occupational and fire risks are subdivided into individual and social risks.

*Individual risks* lead to undesired consequences such as illness, injury, disability, and fatality. *Social risks* occur as a result of emergencies and lead to group injuries, increased mortality, fatalities, and diseases of social groups.

### 6.1.3 Safety Principles and Risks

In order to minimize risks associated with the possibility of injury and fatality on railway transport the URRAN system uses well-known safety principles [3] (MEM, GAMAB, and ALARP) to assess and predict risks.

The MEM (Minimum endogenous mortality) principle is as follows: "the threat associated with the new system should not increase the minimum individual endogenous mortality."

Endogenous mortality is a risk $r$ that takes into account the influence of technological factors on mortality rate of population of a certain age per year. Minimum endogenous mortality rate $R_m$ (the lowest mortality rate for the age group of 5–15 years) in developed countries is assumed as $R_m = 2 * 10^{-4}$ of fatalities/person*year.

When determining the acceptable level of risk according to the MEM principle, the following rule is used: *the danger from a new transport system should not significantly increase the value $R_m$.*

When setting requirements for the minimum tolerable levels of individual and collective (social) risks, one should be guided by the Declaration of the Russian Scientific Society for Risk Analysis "On maximum acceptable levels of risk (MAL)":

- Standard value of the MPL for individual risk per year $R_{\mathrm{IR}} \leq 10^{-6}$
- Standard value of the MPL for the social risk of fatalities of $N$ and more number of people per year: $R_{\mathrm{SR}} \leq 10^{-3}/N^2$ (for newly designed facilities) $R_{\mathrm{SR}} \leq 10^{-2}/N^2$ (for operating facilities).

According to the international agreement, it is generally accepted that a risk of influence of industrial hazards should fall within 10¯7 – 10¯6 (fatalities/person*year), and the value of 10¯6 is the maximum acceptable level of individual risk.

The GAMAB principle (Globalement Au Moins Aussi Bon (France) is generally similar to MEM principle): "*All new transport systems controlled should generally have a degree of risk at least the same as the equivalent existing system.*" This wording takes into account what was achieved and implies the need to improve the system designed through the requirement of "at least." It does not address a specific type of risk, as indicated by the word "generally." Infrastructure and rolling stock suppliers are free to choose between different types of risks inherent to the infrastructure and rolling stock, and take an appropriate approach, i.e. qualitative or quantitative.

The ALARP principle (As Low As Reasonably Practicable): "*The risk is as low as reasonably practicable.*" The acceptable level of risk in accordance with the ALARP principle is the level of risk for which the cost of its achieving is economically efficient.

The essence of the ALARP principle is illustrated in Fig. 6.3. With regard to individual risk, Fig. 6.3a shows three regions: (1) *The region of unacceptable risk*, when a risk must be reduced at any cost; some risks are so great and consequences are so unacceptable that they are unacceptable and cannot be justified in any case. The upper boundary defines the levels of risk that are unacceptable. If the level of risk cannot be lowered below this boundary, then the risk must be excluded; (2) *Region of negligible risk*—no risk reduction measures are required; (3) *ALARP* area—this region between the upper and lower boundaries is called the ALARP region. It should be emphasized that it is not enough to determine that any type of risk is located in the ALARP region. It should be made as low as practicable.

There are various ways to apply the ALARP principle. In some cases, it is sufficient to indicate that the best available up-to-date standards and practices were used. In case of new types of activities or doubts about the adequacy of up-to-date standards and practices, the concept of cost–benefit analysis is applied. The content

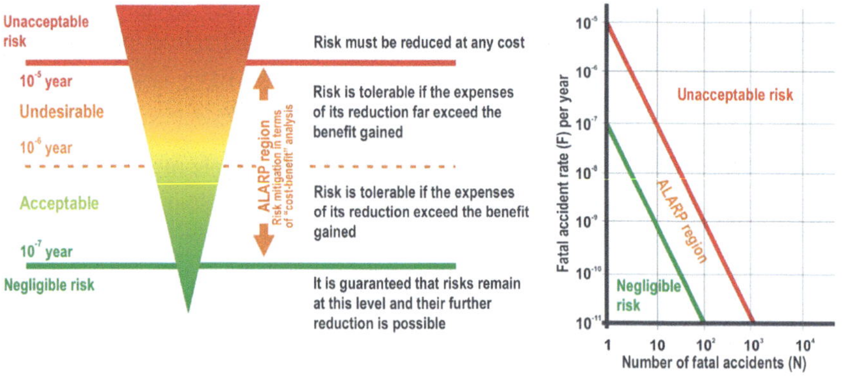


**Fig. 6.3** The essence of the ALARP principle

of this concept is as follows. If the risk of an object is in the ALARP region and its reduction is impossible or the costs of its reduction far exceed the expected benefits, then such a risk is *undesirable* but allowed (Fig. 6.3). Here the last word remains with the operating organization. The lower part of the ALARP region corresponds to a situation where the disproportion between costs of risk reduction activity in question and benefits does not exceed the specified value. In these cases, funds should be spent on risk reduction. This risk is usually called acceptable. The results of a cost–benefit analysis often depend on the way the consequences of a hazardous event are assessed (for example, the value of human life or deaths averted). The use of the ALARP principle is illustrated in GOST 33433-2015. "Functional safety. Risk management on railway transport."

## 6.2 Methodology of the Risk Management Process

When solving complex safety issues, the risk management process methodology is applied. It provides for making decisions based on determining the frequency (probability) of undesirable events occurrence and the size of their consequences.

The main task of risk management in railway transport is to achieve and maintain an acceptable level of functional safety of infrastructure facilities and rolling stock of railway transport.

Based on the analysis of findings in the field of risk management and taking into account current requirements, the methodology of risk management in railway transport should meet the following basic principles:

- A decision associated with a risk shall be cost-effective and shall not generate a negative impact on the results of financial and operational activities

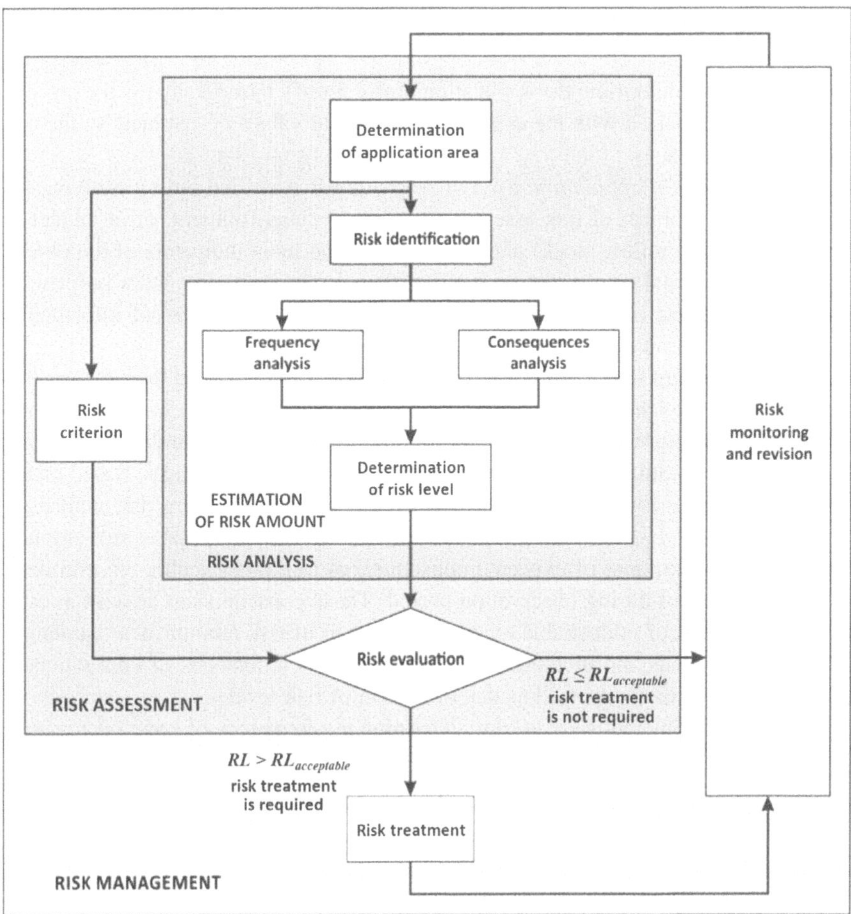


**Fig. 6.4** Risk management process and its stages

- Risk management shall be carried out within the framework of the corporate strategy (for example, the strategy of the Russian railways holding)
- When managing risks decisions made should be based on the necessary amount of reliable information.

The organization of the risk management process is based on the sequential execution of individual stages, procedures and steps using known approaches, methods and technologies that are applied in practice, available information about external and internal conditions and about the facility or process associated with the risk to be managed.

Figure 6.4 (see [4]) shows the subdivision of the risk management process into separate stages.

### 6.2.1 Risk Assessment Stage

At this stage, we determine the application area, identify hazards, assess the amount of risk, and compare it with the established threshold values determined by the risk tolerability criteria.

*Determination of application area of risk analysis* is carried out by studying the properties of an object of risk assessment (a person, environment, or an object of infrastructure and rolling stock) and by analyzing the main indicators of the object, internal and external conditions, as well as other initial data. The main purpose of studying the object of risk assessment is to determine the sources of information about this object and methods of using this information.

*When identifying a risk,* we detect a list of undesired events and their factors that can cause harm to a human, deterioration of environment, damage to the infrastructure and rolling stock. The result of risk identification is a list of undesirable events and their factors leading to an accident or other negative consequences. Based on the results of risk identification, we determine further actions regarding risk analysis.

*Estimation of risk amount is* determination of the risk (measure) amount, expressed in the amount of expected consequences to human health, environment, and material values during observation period. These consequences arise as a result of the occurrence of undesirable events. Estimation of risk amount is a mandatory part of risk analysis and includes analysis of frequency, analysis of consequences and their combinations, as well as determination of risk level.

The following methods are used to determine *the frequency of event occurrence*:

- Assessment of frequency of occurrence of this event in past on the basis of statistical data (data accumulated over a certain period of operation of an object of infrastructure or rolling stock in question, statistical data on traffic accidents and other events, etc.) and forecasting the frequency with which this event will occur in future
- Assessment of frequency of this event occurrence on the basis of data on technical equipment failures occurred over a certain period of time per unit of operational work (for each railway transport facility)
- Forecasting the events occurrence frequencies using methods of analyzing the diagram of possible failures of an infrastructure facility or rolling stock, as well as using data science technology
- Assessment based on expert opinion. When conducting expert assessments, one should take into account any available information about an infrastructure facility or rolling stock (statistical, experimental, constructive data, etc.). There are methods for obtaining expert assessments that eliminate ambiguity, for example, the Delphi method, pair matching, classifications of risk groups, and others.

*Consequence analysis* can be performed both in the form of a simple description of the results using simplified analytical methods, and in the form of detailed quantitative modeling (for example, using computer simulation models). When analyzing consequences, we choose an undesirable event on the basis of the results of risk identification and a description of all consequences arising from this event, including:

- The consequences that caused damage to the considered object of risk assessment
- The consequences that may appear after a certain period of time (if their consideration does not go beyond the application area of the risk analysis)
- Secondary consequences affecting adjacent infrastructure and rolling stock.

The size of consequences can be set both qualitatively and quantitatively (quantitative assessment is the most objective).

If quantitative estimates of the frequency and size of consequences are available, a quantitative evaluation of risk level is carried out, using various mathematical formulations.

In general, *the determination of risk level R* provides for the expression of risk using two values: the frequency or probability $P$ of an undesirable event and size of its consequences $C$. If the consequences are unknown, a risk can be expressed only by probability or frequency. We can also define a risk as the probability of exceeding a certain limit (threshold) by a certain random parameter. Risk is often measured in monetary terms if there is a threat to material assets. If various consequences of an undesirable event are the same or very large, then it is sufficient to consider only the corresponding frequencies or probabilities in order to compare risks. In case of a health risk, the consequences can be partially quantified for such categories as downtime or costs of replacement of personnel, insurance payments, etc. If there is a risk associated with a fatality, quantitative estimates of the consequences are not available in most cases.

*Tolerable risk criteria* determine the acceptable level of risk and are set depending on the methods of risk analysis, the availability of the necessary information, the capabilities and objectives of analysis. The criteria of tolerable risk can be set by regulatory documents, determined when planning a risk analysis, determined in the process of obtaining the results of risk analysis. When performing a risk analysis the main requirements to the choice of the tolerable risk criterion are justification and definite. In general, the basis for determining the tolerable risk should be: the legislation of the Russian Federation, safety rules and regulations in the sphere analyzed, additional requirements of the authorized bodies in the field of safety, information about emergency events occurred and their consequences, experience in this type of activity. Tolerable risk criteria are usually based on operational, technical, economic, regulatory, social or environmental factors, or its combination.

The basic principles of risk acceptance (ALARP, MEM, and GAMAB) are discussed in item 6.1.3. In general, an acceptable risk represents some kind of compromise between an acceptable level of safety and the economic potential for achieving it. The level of negligible risk limit is usually set as 0.01 of the maximum acceptable risk limit (that is, values less than $10^{-8}$ are considered negligible for the above-mentioned tolerable individual risk level $10^{-6}$).

## 6.2.2 Risk Evaluation and Treatment Stage

*Risk evaluation (or comparison)* follows the risk analysis and completes the risk assessment procedure. When evaluating a risk, the obtained assessment of the risk level $R$ is correlated with one (acceptable level ($RL_{\text{acceptable}}$)) or several specified risk levels, which are determined on the basis of the acceptable risk level. The acceptable risk level is determined by the criteria for tolerable risk. The results of risk evaluation are typically presented using a risk matrix. Description of risk matrices is given in Sect. 6.3. Based on the results of risk evaluation, we make a decision whether risk treatment is necessary and a decision on priority order of risk treatment. Risk treatment decisions are made by company's management on the basis of reputational, legal, financial, and other considerations, including risk perception.

In practice, the following four options for *risk treatment* measures are widely applied:

- *Risk prevention* is consideration of ways to eliminate a hazard or change the process, activity in such a way that a hazard relating to them no longer arises. When identified risks are considered to be too high, one could make a decision to completely terminate planned or existing activity
- *Risk transfer* is transfer of a risk to a third party that can take risk, for example, an insurance company or through the transfer of responsibility from the operating organization to suppliers or safety management services, as well as outsourcing. Risk transfer can cause new risks or modify existing identified risks, therefore it may be necessary to identify new risks or revise existing risks
- *Risk reduction* is the use of appropriate means of monitoring dangerous failures and other undesirable events and implementation of technical and organizational activities to reduce the frequency (probability) of an undesirable event or the size of its possible consequences. Each control means or action can provide one or more types of protection: prevention, containment, detection, reduction, recovery, correction, monitoring, and reporting
- *Risk acceptance* is making a decision not to treat a risk. The organization should come to a decision on risk acceptance on the basis of the acceptance criteria:
  - Successful risk reduction when the residual risk meets the risk tolerability criteria after the implementation of the measures
  - Risk retention, that is, even if the initial or residual risk exceeds the tolerability criteria, the management of the organization makes a decision to accept the risk, taking into consideration various conditions such as budget, time constraints, etc.

Risk treatment should be clearly ranged by priority. And individual risk treatment will be implemented for each risk. Prioritization can be done using a variety of techniques, including risk ranking and cost–benefit analysis. Upon treating risk some residual risk remains. As a rule, it is required to assess residual risk in order to make a decision to accept it.

### 6.2.3 Risk Monitoring and Review Stage

***Risk Monitoring*** The risks are not static. Hazards, their occurrence, frequency of occurrence and consequences can change significantly under the influence of various factors. To identify and control these changes risk monitoring is required at various stages of the risk management process. The result of risk monitoring can be a recommendation to review a risk.

The main purpose of risk monitoring is to reduce the uncertainty when assessing a risk. When making decisions on risk management, the use of information about risk introduces risk monitoring into risk management loop, which makes it possible to consider the risk monitoring process as an integral part of the risk management decision-making process.

Risk monitoring tasks can be roughly divided into three main groups:

- Analytical monitoring tasks (include identification of risk signs, estimation of risk amount, risk evaluation)
- Situational monitoring tasks (include monitoring sources of information about risk, monitoring the dynamics of risk, control of parameters affecting the risk)
- Operational monitoring tasks include control of risk management results.

***Risk Review*** If we do not take special measures, then predicting the level of risk based on the previously obtained risk assessments will be erroneous at some point, since these assessments do not take into account the occurred changes of influencing factors. For example, new hazards, changes in their frequency (probability) of occurrence, or the size of consequences can significantly increase risks previously assessed as negligible.

Risk review is carried out:

- In case of identifying any changes associated with hazards, their factors and occurrence, frequency of occurrence and the size of the consequences
- In case of significant changes in organizational, operational, technical, economic, regulatory, social, environmental, and other factors affecting the functioning of railway transport
- Periodically, after a specified time interval since the latest review (time interval is determined separately for each type of risk). When reviewing risks, it is advisable to consider each risk both separately and in aggregate, since their aggregation during the assessment also sums up the amount of potential losses. If the risks do not fall into negligible or acceptable risk category, they should be treated using one or more risk treatment measures. Changes in factors (affecting the frequency (probability) and consequences of an undesirable event) may also affect on possibility of taking various risk treatment measures or on their cost. Therefore, risk review activities should be periodically repeated. Chosen risk treatment measures should also be reviewed if necessary.

## 6.3 Risk Matrix

The results of risk estimation are typically represented using a risk matrix. The risk matrix is a modified form of the risk graph and allows representing the risk levels in "frequency-consequences" coordinates, specified both qualitatively and quantitatively.

Since the risk level $R$ is expressed by the product of the frequency $f$ of the occurrence of an undesirable event by its specific damage C, the scales of the frequency and the scale of consequences on the risk graph should be logarithmic. This ensures that hyperbolic dependences $f = R/C$ ($R$ = const) are represented in the coordinates "frequency-consequences" in the form of straight lines and allow moving from the risk graph to the risk matrix with the least significant loss of accuracy.

If a straight line $f = R_{\text{acceptable}}/C$ is given, where $R_{\text{acceptable}}$ is established acceptable level of risk, then all points lying below this straight line will correspond to the less acceptable level of risk, and those located above this line will correspond to the more than acceptable level of risk.

For the practical use, as a rule, several interval ranges are specified for assessing risk in several categories. So, the area in logarithmic coordinates "frequency-consequences," limited by two straight lines $f = R_1/C$ and $f = R_2/C$ ($R_1 < R_2$), will correspond to the set of points with risk level values from $R_1$ to $R_2$.

Figure 6.5 shows the typical form of the risk matrix with the literal notation of the scale levels. In place of the literal notation, there should be numerical values

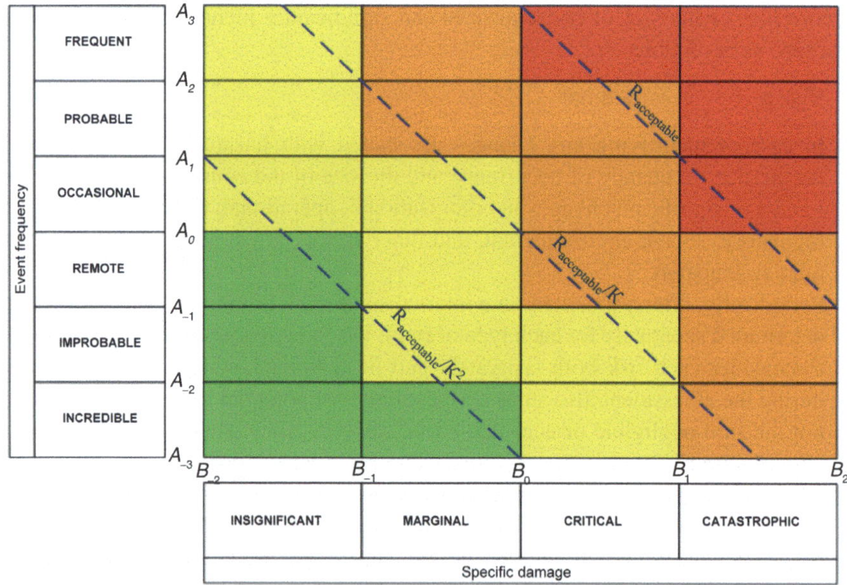


**Fig. 6.5** Standard risk matrix form

| Risk level | Range of values |
|---|---|
| Unacceptable | $R > R_{acceptable}$ |
| Undesired | $0.1 \cdot R_{acceptable} \leq R < R_{acceptable}$ |
| Tolerable | $0.01 \cdot R_{acceptable} \leq R < 0.1 \cdot R_{acceptable}$ |
| Negligible | $R < 0.01 \cdot R_{acceptable}$ |

**Fig. 6.6** Estimated scale of risk

obtained when binding the scales to absolute values. The figure also shows an auxiliary line $f = R_{\text{acceptable}}/C$ for clarification purposes. This line illustrates the reflection of one scale to another based on the acceptable level of risks (when constructing a report risk matrix, there should be no auxiliary line), and the recommended ratio of the cell sides (height to width) is 1:2. For ease of fitting on an A4 sheet, the width of the cell can be selected 30 mm.

**Choice of Risk Scale Parameters**

The first task in constructing a risk matrix is to select the parameters of the risk scale. This issue was solved in works [5, 6 and others]. Since quantitative risk assessment is of greatest interest, such parameters will be:

- The number of estimated intervals of the risk scale
- The response of the risk scale, expressed by the relative spacing $K$ of the risk scale
- Bind value—the absolute value of risk corresponding to a given point on the risk scale; the value of $R_{\text{acceptable}}$ is usually set at the point of the scale corresponding to the acceptable level of risk.

As a rule, the number of risk intervals (categories) is specified for a wide range of risks and is not subject to change in the process of constructing risk matrices. For all considered risks, we adopt a risk scale, which has four estimated intervals of risk values (Fig. 6.6).

The bind value is uniquely determined by the acceptable risk level, which is set for each risk type, or by two ALARP levels, which are also set for each risk type.

It should be noted that if two ALARP levels are set (acceptable and negligible), then the response of the risk scale is uniquely determined:

$$K = \sqrt{\frac{R_{\text{acceptable}}}{R_{\text{negligible}}}} \tag{6.1}$$

where $R_{\text{acceptable}}$ is the acceptable level of risk;

$R_{\text{negligible}}$ is the negligible level of risk [7].

Based on the recommended values $K$ defined above, when setting two levels of risk according to ALARP, it is advisable to keep their ratio close to $K^2$, where $K$ is selected from row 2; 2.5; 3; 4; 5; 6; 7; 8; 10; 14.

If only $R_{acceptable}$ is set, then the response $K$ of the risk scale is chosen on the basis of the required range of risk values and the analysis of the statistics of frequencies and consequences over an extended observation interval (which will be discussed below).

**Choice of Parameters of Scales of Frequencies and Consequences**
By analogy with risk scale, we should also specify response, number of intervals, and bind value for the frequency scale and the consequence scale.

In this case, the response $k_f$ of the frequency scale and $k_c$ of the consequences scale are related to the response $K$ of the risk scale as follows [8]:

$$K = k_f k_c = K^{\alpha} K^{\beta}, \alpha + \beta = 1 \quad (6.2)$$

where $\alpha$ and $\beta$ are the indicators of the scales of frequencies and consequences, respectively.

It was found that in order to ensure the highest reliability of the risk assessment results one should set $\alpha = 1/3$, $\beta = 2/3$ when transiting to the risk matrix.

In order to unify the form of the risk matrix, it is advisable to set the standard numbers of intervals for the frequency scale and the scale of consequences. Taking into account the above optimum values of the indicators $\alpha$ and $\beta$, it is recommended to use the following risk matrix as a standard form: a risk matrix with the number of intervals (cells) of the frequency scale $m = 6$ and the number of intervals (cells) of the scale of consequences $n = 4$. Then the issue of choosing the parameters of the risk matrix for a given acceptable level of risk will be reduced to the choice of the most appropriate value K and specifying the bind value for one of the scales (frequencies or consequences). These parameters can be determined on the basis of statistical data on the frequency of occurrence of an undesirable event or its consequences over the previous whole observation period available.

For this purpose, one should estimate the relative ranges of variation of frequency and specific consequences. Then an appropriate $K$ value can be chosen on the basis of the corresponding nearest largest value of range of scale of frequencies and the consequences for the accepted standard $6 \times 4$ matrix form.

The ranges of matrix scales should be chosen taking into account the necessary safety factors.

The binding of the frequency scale of the matrix is as follows.

Required range of frequency value:

$$A = \frac{a_{max} a_{min} f_{max}}{f_{min}} \quad (6.3)$$

where:$a_{max} = 2$ and $a_{min} = 1{,}5$ are the reserve factors for the upper and lower limits of the frequency range, respectively.

$f_{min}$ and $f_{max}$ are the minimum and maximum value of the frequency of occurrence of an undesirable event in a sample (for an extended observation interval), 1/year.

Bind value is the center of the frequency scale:

$$A_0 = \sqrt{\frac{a_{max} f_{max} f_{min}}{a_{min}}} \tag{6.4}$$

Similarly, the required range of the scale of consequences is determined for the scale of consequences:

$$B = \frac{b_{max} b_{min} C_{max}}{C_{min}} \tag{6.5}$$

where:$b_{max} = 2$ and $b_{min} = 1{,}5$ are the reserve factors for the upper and lower limits of the range of consequences, respectively.$C_{min}$ and $C_{max}$ are the minimum and maximum value of the specific consequences of an undesirable event in the sample, train-hours (hours, units).

Bind value is the center of the consequences scale:

$$B_0 = \sqrt{\frac{b_{max} C_{max} C_{min}}{b_{min}}} \tag{6.6}$$

For the obtained value *A* (row 3 in Table 6.1), we find the nearest larger value and fix the corresponding value *K*.

Similarly, for the obtained value *B* (row 4 in Table 6.1), we find the nearest larger value and fix the corresponding value *K*.

We select the larger of the two fixed *K* values. Also we select the scale for which the *K* value was the larger of the two values. For this scale, first we bind to absolute values.

If both *K* values are the same, then one select the scale in which the ratio $A/A_T$ or $B/B_T$ is the largest.

Let us consider two options for binding scales of the risk matrix.

***Option 1*** Binding on the frequency scale. In this case, the center of the frequency scale is assigned the value $A_0$ calculated above.

Then we perform the "projection" of the frequency scale onto the scale of consequences and check the compliance of the latter with the required range of values:

- the projected values of $B_{-2}$ and $B_2$ of the scale of consequences are calculated based on the obtained value $A_0$ and the acceptable level of risks:

$$B_{-2} = \frac{R_{acceptable}}{A_0} K^{(-2\beta-1)}; B_2 = \frac{R_{acceptable}}{A_0} K^{(2\beta-1)} \tag{6.7}$$

- we check whether the inequation condition is met

**Table 6.1** Parameters of the 6 × 4 risk matrix scales

| Indicators | Risk scale spacing, $K$ | | | | | | | | | |
|---|---|---|---|---|---|---|---|---|---|---|
| | 2 | 2.5 | 3 | 4 | 5 | 6 | 7 | 8 | 10 | 14 |
| 1. Frequency scale spacing, $k_f = K^{\alpha}$ | 1.26 | 1.357 | 1.442 | 1.587 | 1.71 | 1.817 | 1.91 | 2 | 2.154 | 2,41 |
| 2. Consequence scale spacing, $k_c = K^{\beta}$ | 1.587 | 1.842 | 2.08 | 2.52 | 2.924 | 3.302 | 3.659 | 4 | 4.642 | 5.809 |
| 3. Frequency range, $A_T$ ($m = 6$) | 4 | 6.25 | 9 | 16 | 25 | 36 | 49 | 64 | 100 | 196 |
| 4. Consequences range $B_T$ ($n = 4$) | 6.35 | 11.51 | 18.72 | 40.32 | 73.1 | 118.9 | 179.3 | 256 | 464.2 | 1138.6 |

$$\frac{C_{\min}}{b_{\min}} \geq B_{-2}. \tag{6.8}$$

If this inequation is *not met* (this means that consequences scale of the matrix is shifted up), then we fix the lower limit of the frequency range (taking into account reserve) $A_{-3} = \frac{f_{\min}}{a_{\min}}$ and select such a coefficient $K$, for which

$$\frac{C_{\min}}{b_{\min}} \geq B_{-2} = \frac{R_{\text{acceptable}}}{A_{-3}} K^{(-3\alpha-2\beta-1)} = \frac{R_{\text{acceptable}} a_{\min}}{f_{\min}} K^{(-3\alpha-2\beta-1)}$$

It follows from this inequation that:

$$K \geq \left( \frac{f_{\min} C_{\min}}{R_{\text{acceptable}} a_{\min} b_{\min}} \right)^{\frac{1}{-3\alpha-2\beta-1}} \tag{6.9}$$

We selected the largest $K$ (from Table 6.1) that satisfies this condition. Next, we calculate levels of the frequency scale on the basis of the already set value $A\_3$:

$$A_j = A_{-3}(K^{\alpha})^{3+j} \tag{6.10}$$

where $j = -2, -1, 0, 1, 2, 3$ is the conditional number of level between the cells of the matrix;

- if $\frac{C_{\min}}{b_{\min}} \geq B_{-2}$, then the following inequation is checked:

$$C_{\max} b_{\max} \leq B_2 \tag{6.11}$$

If this inequation *is not met* (this means that consequences scale of matrix is shifted down), then we fix upper limit of the frequency range (taking into account reserve) $A_3 = f_{\max} a_{\max}$ and select such a coefficient $K$, for which

$$C_{\max} b_{\max} \leq B_2 = \frac{R_{\text{acceptable}}}{A_3} K^{(3\alpha+2\beta-1)} = \frac{R_{\text{acceptable}}}{f_{\max} a_{\max}} K^{(3\alpha+2\beta-1)}$$

This inequation implies:

$$K \geq \left( \frac{f_{\max} C_{\max} a_{\max} b_{\max}}{R_{\text{acceptable}}} \right)^{\frac{1}{3\alpha+2\beta-1}} \tag{6.12}$$

We selected the smallest $K$ (from Table 6.1) that satisfies this condition. Next, the levels of the frequency scale are calculated on the basis of the already set value $A_3$:

$$A_j = A_3(K^{\alpha})^{j-3} \tag{6.13}$$

where $j = -3, -2, -1, 0, 1, 2, 3$ is the conditional number of the level between the cells of the matrix;

- if $\frac{C_{\min}}{b_{\min}} \geq B_{-2}$ and $C_{\max}b_{\max} \leq B_2$, then subsequent calculations are carried out with the coefficient $K$, selected initially; in this case, the frequency scale is formed on the basis of the value $A_0$ obtained above by formula (6.4), as follows:

$$A_j = A_0(K^{\alpha})^j \tag{6.14}$$

where $j = -3, -2, -1, 1, 2, 3$ is the conditional number of the level between the cells of the matrix;

- the projection of the center of consequences scale is calculated based on the value $A_0$:

$$B_0 = \frac{R_{\text{acceptable}}}{A_0 K} \tag{6.15}$$

- the levels of the scale of consequences are calculated based on set value $B_0$:

$$B_j = B_0\left(K^{\beta}\right)^j \tag{6.16}$$

where $j = -2, -1, 1, 2$ is the conditional number of the level between the cells of the matrix.

***Option 2*** Binding on a consequences scale. In this case, the center of the consequences scale is assigned the $B_0$ value calculated above.

Then we perform the "projection" of the consequences scale onto the frequency scale and check the compliance of the latter with the required range of values. This is done as follows for a standard risk matrix form:

- The projected values $A_{-3}$ and $A_3$ of the frequency scale are calculated based on the obtained value $B_0$ and the acceptable level of risks:

$$A_{-3} = \frac{R_{\text{acceptable}}}{B_0} K^{(-3\alpha-1)}; A_3 = \frac{R_{\text{acceptable}}}{B_0} K^{(3\alpha-1)}. \tag{6.17}$$

- We check whether the inequation condition is met

$$\frac{f_{\min}}{a_{\min}} \geq A_{-3} \tag{6.18}$$

If this inequation is *not met* (this means that the frequency scale of the matrix is shifted up), then we fix the lower limit of the consequences range (taking into account reserve) $B_{-2} = \frac{C_{\min}}{b_{\min}}$ and select such a coefficient $K$ for which

$$\frac{f_{\min}}{a_{\min}} \geq A_{-3} = \frac{R_{\text{acceptable}}}{B_{-2}} K^{(-3\alpha-2\beta-1)} = \frac{R_{\text{acceptable}} b_{\min}}{C_{\min}} K^{(-3\alpha-2\beta-1)}$$

Condition (6.19) follows from this inequation.

$$\begin{cases} K \geq \left( \frac{f_{\max} C_{\max} a_{\max} b_{\max}}{R_{\text{acceptable}}} \right)^{\frac{1}{2\alpha+\beta}} \\ K \geq \left( \frac{f_{\min} C_{\min}}{R_{\text{acceptable}} a_{\min} b_{\min}} \right)^{\frac{1}{-4\alpha-3\beta}} \end{cases} \tag{6.19}$$

We select the largest value $K$ (from Table 6.1) that satisfies (6.19). Then we calculated levels of consequence scale on the basis of already set value $B_{-2}$:

$$B_j = B_{-2} \left( K^{\beta} \right)^{2+j} \tag{6.20}$$

where $j = -1, 0, 1, 2$ is the conditional number of level between the matrix cells.

- if $\frac{f_{\min}}{a_{\min}} \geq A_{-3}$, then the following inequation is checked:

$$f_{\max} a_{\max} \leq A_3 \tag{6.21}$$

If this inequation is *not met* (this means that the frequency scale of the matrix is shifted down), then we fix upper limit of the consequences range (taking into account reserve) $B_2 = C_{\max} b_{\max}$ and select such a coefficient $K$ for which

$$f_{\max} a_{\max} \leq A_3 = \frac{R_{\text{acceptable}}}{B_2} K^{(3\alpha+2\beta-1)} = \frac{R_{\text{acceptable}}}{C_{\max} b_{\max}} K^{(3\alpha+2\beta-1)}.$$

Condition (6.13) follows from this inequation. We select the smallest $K$ (from Table 6.1) that satisfies condition (6.13). Then we calculated levels of consequence scale on the basis of already set $B_2$ value:

$$B_j = B_2 \left( K^{\beta} \right)^{j-2} \tag{6.22}$$

where $j = -2, -1, 0, 1$ is the conditional number of the level between the cells of the matrix;

- If $\frac{f_{\min}}{a_{\min}} \geq A_{-3}$ and $f_{\max} a_{\max} \leq A_3$, then subsequent calculations are carried out with the coefficient $K$, selected initially; wherein the consequences scale is formed on the basis of the value $B_0$ obtained above by formula (6.7), as follows:

$$B_j = B_0 \left(K^{\beta}\right)^j \tag{6.23}$$

where $j = -2, -1, 1, 2$ is the conditional number of the level between the cells of the matrix;

- The projection of the center of the frequency scale is calculated based on the value of $B_0$:

$$A_0 = \frac{R_{\text{acceptable}}}{B_0 K} \tag{6.24}$$

- the levels of the frequency scale are calculated based on the set value $A_0$:

$$A_j = A_0 (K^{\alpha})^j \tag{6.25}$$

where $j = -3, -2, -1, 1, 2, 3$ is the conditional number of the level between the matrix cells.

The calculated (according to one of the options considered above) levels of frequencies and consequences are placed on the matrix scales (see Fig. 6.5). The result is a matrix suitable for representing a given type of risk.

The value of the frequency $f_i$ of the occurrence of an undesirable event is correlated with the frequency scale of the risk matrix for a given type of risk over the chosen observation interval; we determine a matrix row with a frequency range that includes the available frequency value.

We correlate the specific amount of damage $C_i$ from this event with the damage scale; we determine the column of the risk matrix with the range of damages which includes the existing damage. The cell indicating the estimated risk is defined at the intersection of the row and column data. The color of this cell indicates the risk category.

Even if applying the maximum of the recommended values $K$, the frequency or specific damage from the risk in question falls outside the cells of the risk matrix, then the following options for solving this issue are possible:

- Using one of the shifted matrices instead of the standard one
- Using one of the extended matrices instead of the standard one.

When using shifted risk matrices, we should observe the "poor compatibility" condition, according to which the matrix must contain at least one cell with the "negligible" level and at least one cell with the "unacceptable" level.

Frequency and consequences levels for non-standard matrices are calculated in the same way as for standard matrices.

## 6.4 Assessment of Integral Risk

A measure of the safety of a system's object can be the value of an associated risk which is based on the risks of its constituent factors (elements). Need for definition of the integral risk of an object and a system is as follows. Summing up of risks of all elements is not acceptable, since they may have, for example, different measures (the number of fatalities during a certain period of time is a social risk, and the cost of losses is an economic one). We need some other methodological tool that can transform different measures of safety of objects (elements) into a certain single integral measure of a system's risk. Such tasks occur in medicine, food industry, in transport sector, etc. [9].

Let system A generally consists of a finite set of diverse elements $A = \{a_1, a_1, \ldots, a_i, \ldots, a_j, \ldots, a_k\}$. And there may be a possibility of equivalence of separate constituent elements $a_i \Leftrightarrow a_j$. Safe operation of each system element is estimated by a certain risk value $a_i \longrightarrow R_i$. Risks are formalized with the use of a risk matrix tool. Generally, a risk matrix contains m rows and n columns. Each row corresponds to a certain frequency of a hazard. Columns correspond to possible consequences (damage) $c_1, c_1, \ldots, c_n$. A measure of consequences depends on the object of analysis. It could be a price (in relation to economic, technical, or anthropogenic risks), fatality in relation to social risks, number of negative consequences due to a hazardous event (in relation to moral risks), etc. It is supposed that the frequencies of hazardous events and their consequences are estimated by a posteriori data. This makes it possible to determine the safety violation risks of all system elements at the intersection of the corresponding rows and columns.

Risks for diverse elements are not equal among themselves, for example, $R_1 \neq R_i$ (risks of equivalent elements are equal $R_1 = R_j$). The task is to assess the level of risk of a system based on the results of assessment of risks of its constituent diverse elements. Risks of the elements are supposed to be mutually independent.

In many cases, the system under study consists of diverse objects that differ in scales of consequences and types of risks (for example, technological or social ones). At present one can neither sum up the risks of constituent objects, nor form a common scale of consequences. To assess a system's risk by the risks of constituent diverse elements, it is necessary to have at least one common measure for all risks. If to consider risks in reference to the scales of measurement $f$ and $c$, such common measure is not available. It also applies to the rates of hazardous events that can be many times different for elements $a_i$ and $a_j$. If we consider the risks in regard to the scales of measurements $f$ and $c$, then this common measure is absent. The measure of damage can be different. This also applies to the frequencies of occurrence of hazardous events, which for elements $a_i$ and $a_j$ can be many times different.

However, upon close examination of the constructed risk matrices for the system elements, we find a common measure of risk assessment that is contained in the levels of decision-making.

According to ALARP principle, there are four levels of risk severities. A common field for combining the results is the colors of decisions (risk levels) for each of the

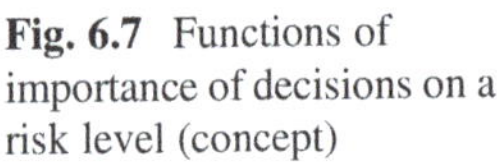
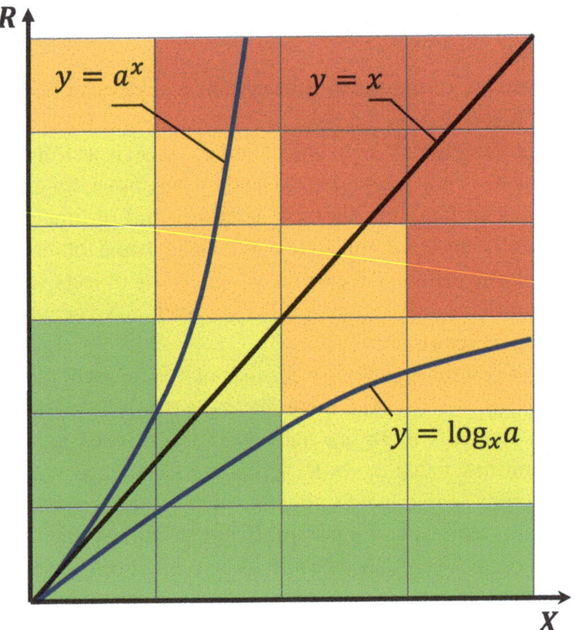


**Fig. 6.7** Functions of importance of decisions on a risk level (concept)

objects. These levels are marked with green, yellow, orange, and finally with red color as their importance grows. The green color of a decision means that risk is so negligible that it can be discounted. The function of importance in green cells of a matrix should have low values (from zero up to a certain insignificant value). In addition to that, the orange color and especially the red color mean the highest degree of severity, and the function of importance in these matrix cells should have the maximum high values. There are three strategies to construct the functions of the importance of decisions on a risk level in accordance with the accepted colors: 1—linear; 2—power; 3—logarithmic. Figure 6.7 provides a conceptual representation of the functions of importance. Thus, to digitize the results of object risk assessment expressed by one of four colors, it is advisable to apply power function (Fig. 6.7).

Strategy 2 specifies a responsible attitude to a change of importance of a decision color. Strategy 3 should be considered as an irresponsible attitude to a decision taken on an object's risk level, as in this case the function of importance mitigates a severity degree of red color that reflects an intolerable risk level.

Figure 6.8 shows stepwise functions of significance with the above indicated bases with four integer values of the degree of a function ($n = 0, 1, 2, 3$).

Step-by-step significance functions for risk level decisions with base $1 \leq a < 2$ do not provide a quick response to a change of the importance of a decision color (Fig. 6.8), especially in the field of high risk levels. However, with base $a > 2$ there is an unreasonably quick response to an undesirable level and especially to an intolerable risk level and almost a neglect of the importance of a tolerable risk level (Fig. 6.8).

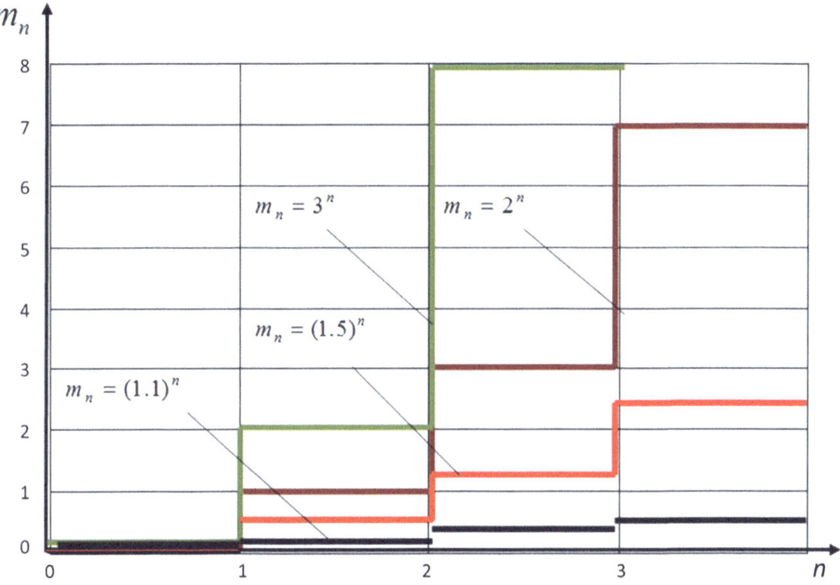


**Fig. 6.8** Stepwise functions of the significance of the colors of decisions about the level of risk

A compromise solution is to choose base 2 of the step function of importance of colors of decisions taken on a risk level.

A color weight is generally defined by the formula (6.4)

$$w_n = \frac{m_n}{\sum_{n=0}^{3} m_n} \tag{6.26}$$

where a digitized value of a risk level color, for instance, *negligible* (the weight graph is marked with green color in Fig. 6.8), with importance function $m_n = 2^n$ is equal to: $m_0 = 2^0 = 1$.

Integral assessment of system risk can be calculated by the formula (6.5)

$$R = \frac{\sum_{n=0}^{3} k_n m_n w_n}{\sum_{n=0}^{3} k_n m_n} \tag{6.27}$$

where:

$k_n$ is a number of system elements with a risk level of the $n$th coloration
$m_n$ is a function of the importance of risk coloration.

Figure 6.9 summarizes weights of colors of typical risk matrix cells as well as values of assessment of decisions taken on integral risk.

| | Index, n | Digitized value, $m_n$ | Color weight, $w_n$ | Values of assessment of decisions taken on integral risk |
|---|---|---|---|---|
| | 0 | 1 | 0.067 | $w_0 \leq R < w_1$ |
| | 1 | 2 | 0.133 | $w_1 \leq R < w_2$ |
| | 2 | 4 | 0.266 | $w_2 \leq R < w_3$ |
| | 3 | 8 | 0.533 | $w_3 = R$ |

**Fig. 6.9** Weights of colors of typical risk matrix cells and values of assessment of decisions taken on integral risk

## 6.5 Technical and Technological Risk Management

Infrastructure facility complexes include track and structure facilities, railway signalling and remote control facilities, railway power supply and electrification facilities, and railway communication facilities. There are *technical risks* associated with these facilities. These risks cause damage to the railway track, damage to signalling equipment, damage to catenary system, traction substations, and power lines, communication system, etc. In many cases, traffic safety on railway transport is determined by *technological risks* that lead to train delays and disruptions to technological processes. There is a typical three-level configuration for organizing the management of technical and technological risks for all complexes of infrastructure facilities. The differences are specific and associated with the special aspects of the organization of work in the units of the infrastructure complexes.

Let us consider the process of risk management using the example of a ***complex of track facilities*** (Fig. 6.10). According to this scheme, risk assessment, including analysis and evaluation is performed at the level of the survey site. Risk assessment includes the stages of risk identification, risk analysis, and risk evaluation (in Fig. 6.10—"constructing a risk matrix").

At the stage of risk identification, the initial data are the data on technical condition of superstructure elements that are collected by crew for inspecting and measuring (lateral and vertical wear of rail lines, malfunctioning switches, parameters of directional error and rail irregularities in profile; track gauge at the derailment point, presence of doted sleepers, sleepers spacing, ballast contamination degree, etc.). This data are updated twice a month. Hazard record book is kept at track section and data from the Railway Diagnostic Center are also used.

Risk analysis, including frequency and consequences analysis, as well as determination of the risk level is performed by personnel appointed by a superintendent.

The final stage of risk assessment is the construction of a risk matrix based on the results of the risk analysis. If the risk does not exceed the acceptable level, then the risk is treated in the normal mode by the engineering section of the track maintenance department (TMDep), then by the engineering section of track maintenance division (TMD) and afterwards by the division of infrastructure territorial directorate. Based on the results of risk monitoring (at the level of track maintenance department and track maintenance division) we review risk, identify it, construct a risk matrix, make a

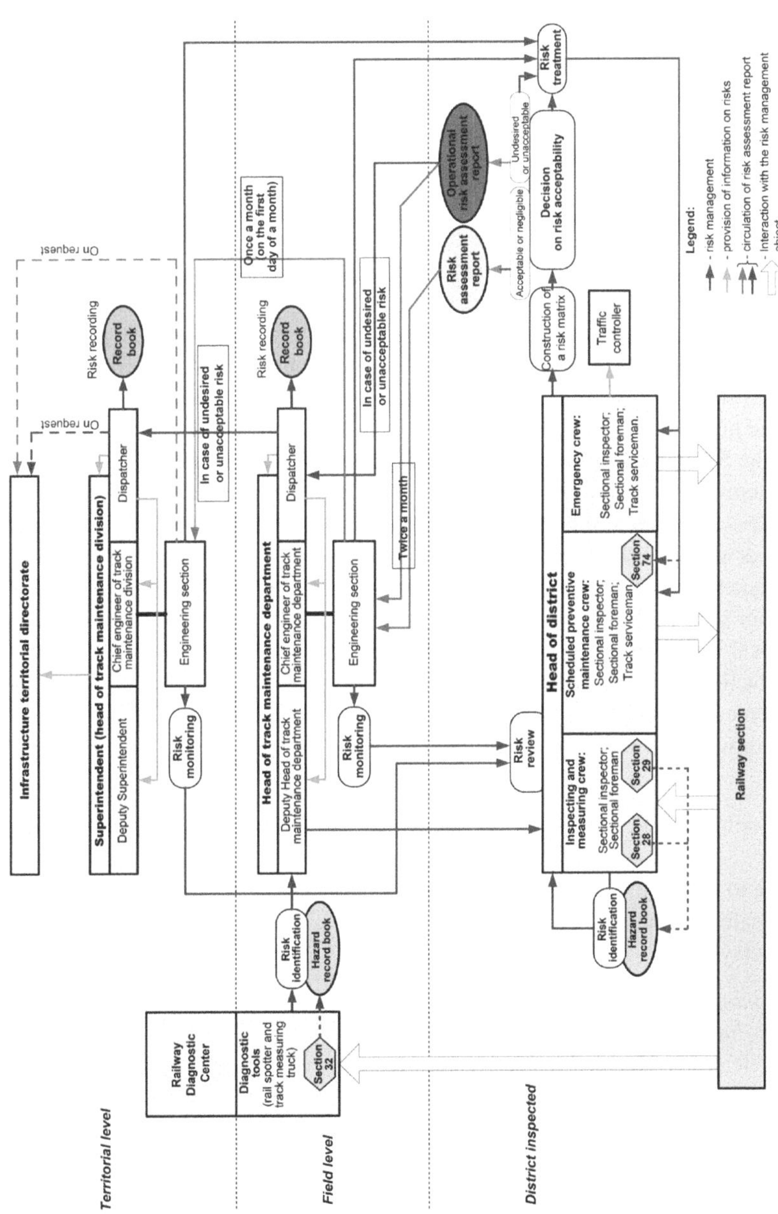


**Fig. 6.10** Organization of management of risks associated with railway track

decision on the acceptability of the risk, its further processing, etc. If the risk exceeds the acceptable level, then one immediately notifies dispatchers of the track maintenance department and track maintenance division hereof, keeps this risk in the record book and promptly takes measures to prevent a traffic accident.

The technical and technological risk management at ***complex of signalling and remote control facilities (SRCF)*** is also organized in three levels. The specific feature of risk management lies in the fact that risk is identified according to the data from technicians, while risk is treated by engineering sections and dispatchers of signalling department (SRCF department) and signalling division (SRCF division).

Risks associated with the functioning of SRCF systems are generally defined as the risks of failures of SRCF systems operated at a given facility and the influence of failures on the process of train movement at a given facility. Risks associated with the functioning of the SRCF systems occur when there is a failure of the SRCF system while SRCF system is operated to manage train movement process. A protective failure of the SRCF system puts the SRCF system into a protective state. SRCF system being in this state can cause various delays in the movement of trains in open lines or station. The number and duration of delays are different for each failure. Dangerous failure of the SRCF system meets one or more dangerous failure criteria and puts the SRCF system into a hazardous state. SRCF system being in a hazardous state can result in the occurrence of injurious effects associated with the movement of trains in open lines or station. It leads to death and serious or moderate harm to human health and direct economic damage. The amount of damage varies for each dangerous failure.

The main risks at ***the complex of electrification and power supply facilities*** are technological and technical risks associated with the functioning of the catenary system, traction and transformer substations, and power supply devices for non-traction consumers.

The following technical or technological risks are associated with catenary system:

- Pantograph damage or damage to other parts of the rolling stock due to wire sagging
- Damage to the rolling stock and/or track superstructure due to the fall of the pillar
- Traffic safety violation
- Train delays.

*Risk management of the catenary system, traction and transformer substations, and power lines* is organized on the basis of a typical three-level configuration. Distinctive feature is that the risks are identified under the supervision of the head of catenary department, head of traction substation, or the head of electrical network department. The initial data are provided by a technician crew twice a month (with the same frequency as at complexes of track facilities and SRCF). Risk assessment reports are submitted to engineering sections and dispatchers of power supply and electrification department (PSE) and Transenergo.

***The risk management process at a complex of communication facilities*** is organized in the same way as at other infrastructure complexes according to a three-level configuration, but it is fundamentally different. Figure 6.11 shows the organization of the management process. According to this scheme, risk is identified

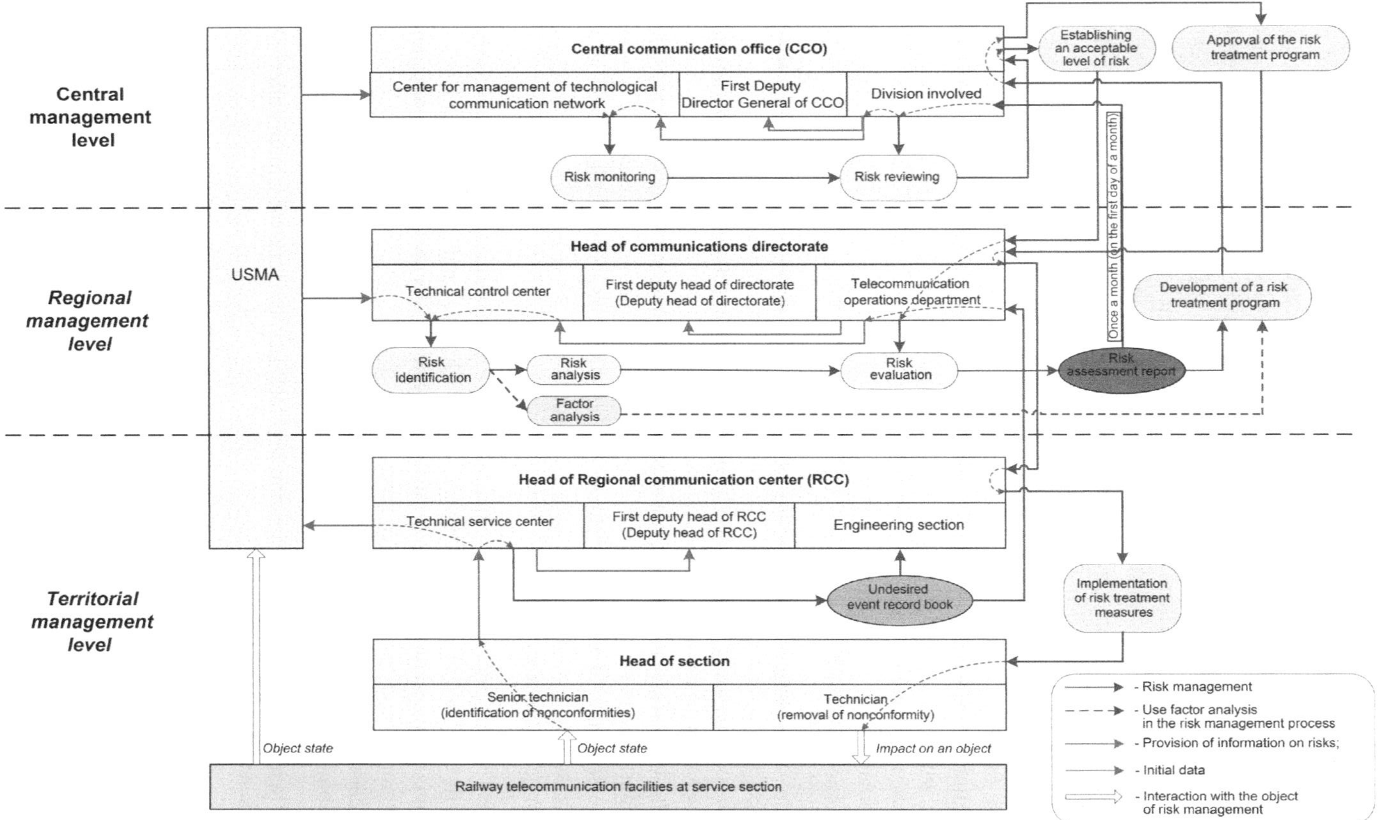


**Fig. 6.11** Organization of risk management related to the functioning of railway telecommunications

at the territorial level of the district inspected. Risk assessment including analysis and evaluation is performed at the regional management level. Acceptable risk levels are established and approved at the central management level. A program of risk treatment including its financing is also approved at central management level.

Risk management stages are performed by personnel appointed by the head of the regional communications center (at the territorial level), the head of the communications directorate (at the regional level), and the chief engineer of the central telecommunications office (at the central management level).

At the stage of risk identification, undesirable events (with an indication of the contributing factors) are kept in a record book. The data of the Unified System for Monitoring and Administrating Communication Networks of the Russian Railways (USMA) are also used. Among undesirable events are:

- Pre-failure of a railway telecommunication facility
- Failure of the railway telecommunication service (failure of category 1 or 2, which led to a violation of the timetable (delay), loss of train-hours and economic losses; as well as failure of category 3)
- Violation of the timetable (delay), which led to the loss of train-hours and economic losses.

One of the reasons for train delays is the failure of the railway telecommunication service, which, in turn, is caused by the failure of technical means.

The complexes of rolling stock facilities include objects of a multiple unit rolling stock (MU) and objects of a locomotive complex.

***The MU facility risk management process*** is organized according to a three-level configuration. In many respects, the structure of this management process is similar to the structure of risk management at track facilities, SRCF, electrification and power supply complexes. At the same time, there are a number of fundamental differences. These include:

- Risk management under the main stages (from identification to risk revision) is performed not at district inspected, but at field level and even regional level
- Initial data on undesirable events are submitted by both rolling stock repair personnel and mu locomotive crews
- Reports on technical and technological risks exceeding tolerable levels are submitted to the repair departments of the mu depot and the regional directorate.

When identifying risks in respect to MU facilities, the following main undesirable events are distinguished:

- Collision or derailment of mu in open lines and stations
- Collision or derailment of the mu during shunting, locomotive servicing, and other operations
- Collision of the mu with motor and tractor vehicles outside level crossings
- Collision of the mu with vehicles and pedestrians at level crossings;
- Train delays, etc.

*The organization of management of risks associated with the operation of locomotives* is almost similar to the risk management scheme in respect to MU. The principal feature of risk management at the locomotive complex is that the service company is involved at all three levels of risk management. If a risk exceeds the acceptable level, the service company is responsible for managing the level of risk.

## 6.6 Fire Risk Management

Fire safety management of both stationary and mobile objects of railway transport covers all stages of the life cycle from their design to decommissioning. The Russian Railways have to simultaneously ensure fire safety of more than 14,000 locomotives, as well as hundreds of stationary facilities where over 300,000 people work. This issue is solved with the help of an automated system for fire risk management. Based on the results of a fire risk forecast this system allows making a decision on whether to repair, replace, or maintain the PIO and the fire safety system.

Two components of methodological tools for risk assessment were defined. The first component uses fire statistics. Its purpose is to assess posterior probabilities of a fire at railway stationary and mobile facilities. The special aspect of assessing the probability of fires (based on fire statistics) is that facilities of each type are divided into groups according to regional criteria. The operating conditions of facilities in groups differ in such parameters as the maximum and minimum external temperature, the quality of repair, the traffic intensity, etc. All of these factors can have an effect on sets of elementary outcomes that contribute to a fire event. Questions arise about the belonging of samples (groups) of the same general population and the need to assess the probability of fire at railway facilities, taking into account the belonging to a particular group of facilities. To answer these questions, hypotheses regarding the equality of sample characteristics are checked using the Pierson's criterion, the Kolmogorov–Smirnov criterion. It was determined that most groups of facilities belong to the same general population. However, it was also determined that locomotives with similar design characteristics but different performances belong to different general populations (2TE10 and 3TE10). Based on the calculated values of the probabilities of fire and known levels of fire consequences, we construct a risk matrix for railway facilities; identify groups of facilities that provide undesired or unacceptable levels of risk.

The second component for processing data on the states of fire safety uses the results of diagnosis of railway facility malfunctions (causing an increase in fire hazard) as initial information. For such facilities, we simulate the sequences of events leading to the occurrence of fire. Decision-making on the order of priority of measures to ensure the fire safety of railway facilities should be based on the hazard assessment of the set of identified states. Fire hazard indicators are: the probability of a fire and the time to the occurrence of a dangerous state. To assess these indicators, one construct a model with the help of which it is possible to represent the process of transition of railway facility into fire hazard state.

A key characteristic of a fire hazard is the probability of fire occurrence, i.e. the probability of transition from a factual non-hazardous state into a given hazardous state according to the model. For a priori assessment of this probability, it is envisaged to identify the factual state of facility malfunction and to assess the possibility of occurrence of a malfunction due to which the object will transit into a fire hazardous state. This procedure is called a fire safety audit. In accordance with standard [10], fire safety audit is categorized as follows: declaration audit, re-audit, and supervisory audit. The declaration audit is the primary fire audit. In case of successful completion of the declaration audit, the fire safety of the facility is declared. A re-audit of facility fire safety is carried out if one fails a declaration fire safety audit or its results are unsatisfactory.

Scenarios of typical fire hazardous events and the facility states preceding them were developed on the basis of statistical data on fires, flame development, and violations of fire safety requirements. Fire hazard classifiers were developed for these scenarios. The classification of fire hazardous states also allows us to distinguish significant violations of fire safety requirements common to railway facilities. Using classifiers of violations of fire safety requirements and fire hazardous states, experts identify fire hazardous states of facilities (fire safety audit).

The fire safety audit is based on the analysis of initial data on fire hazard states of railway facilities and processing of expert opinions on the severity of possible consequences. The need for expert opinion is caused by the fact (among other things) that significant part of the data on the fire hazard states is non-numerical.

As a result of the audit, either a set of fire hazardous states of a facility or a set of violations of fire safety requirements is formed (the choice of an approach depends on the complexity of formation of a particular set). A set of states or events is sufficient data to estimate the prior probability of fire at facility.

With regard to traction rolling stock (TRS), Table 6.2 shows a typical form of data of assessment of priori probability of a component fire, as well as estimated ratios for assessing the probability of a unit fire or/and TRS fire as a whole. The table provides for taking into account the influence of compensating effect at the level of locomotive unit. In the formula expression for calculating the probability of a unit fire, there is a constituent $P_{M/R}^{\text{overmileage}}$ —the probability of an undesirable event. That event can lead to a fire hazardous failure, depending on whether there is locomotive overmileage after the regular maintenance or repair (M/R). This constituent is determined by expert opinion. "Units..." columns of Table 6.2 give the values $\Delta_{M/Rj}$—the coefficient taking into account the share of the *i*th fire hazardous states for the *j*th fire hazard component. These values cannot be detected during a fire audit.

Figure 6.12 shows a matrix of fire risks for a practically established range of diesel locomotives fire probability (from $2.5 \times 10^{-}4$ to $1.2 \times 10^{-}5$). The series of locomotives are also indicated. When constructing the matrix, the costs of M, R (1,2,3), intermediate overhaul (IL), and overhaul (O) were taken into account as damages.

When assessing risks of fire at stationary facilities (for example, passenger buildings), we consider the consequences expressed by the minimum wage (MW). Figure 6.13 shows the results of test calculations of individual and collective risks of

**Table 6.2** Typical form of data on the assessment of the prior probability of TRS component fire

| Component | Unit "A" | | Unit "B" | | Unit "C" | |
|---|---|---|---|---|---|---|
| | $P_j^{\text{component}}$ | $\Delta_{M/R_j}$ | $P_j^{\text{component}}$ | $\Delta_{M/R_j}$ | $P_j^{\text{component}}$ | $\Delta_{M/R_j}$ |
| $P_j^{\text{units}} = \max\left\{P_j^{\text{component}}\right\} + P_{\frac{M}{R}}^{\text{overmileage}} - 2\max\left\{P_j^{\text{component}}\right\} P_{M/R}^{\text{overmileage}}$ | | | | | | |
| $P_{\text{TRS}} = \max\left\{P_j^{\text{units}}\right\}$ | | | | | | |
| $P_{TRS}^{\text{КОМ}} = \max\left\{P_{j\text{ККО}}^{\text{units}}\right\}$ | | | | | | |

| Event | Fire probability | Fire risk level | | | | |
|---|---|---|---|---|---|---|
| Frequent | $P>2.55\cdot10^{-4}$ | Tolerable | Undesired | Unacceptable 3TE10 | Unacceptable | Unacceptable |
| Probable | $1.2\cdot10^{-4}<P\leq2.55\cdot10^{-4}$ | Tolerable | Undesired | Undesired | Unacceptable 2TE116 | Unacceptable |
| Occasional | $5{,}57\cdot10^{-5}<P\leq1.2\cdot10^{-4}$ | Tolerable | Tolerable | Undesir 2TE10 | ceptable | Unacceptable |
| Remote | $2.56\cdot10^{-5}<P\leq5.57\cdot10^{-5}$ | Negligible | Tolerable | 2M62 Undesir TEM2 | esired | Unacceptable |
| Improbable | $1.23\cdot10^{-5}<P\leq2.56\cdot10^{-5}$ | Negligible | Tolerable | TEP70 Tolerable | Undesired | Unacceptable |
| Incredible | $P\leq1.23\cdot10^{-5}$ | Another Negligible | Negligible | ChME3 Tolerabl TEM7 | sired | Undesired |

**Fig. 6.12** Fire risk matrix for diesel locomotives of JSC "Russian Railways" (the central management level)

| Station building | Fire risk | Collective (social) risk |
|---|---|---|
| Samara | $10^{-6}$ | $4*10^{-6}$ |
| Belorussian | $10^{-6}$ | $5.3*10^{-6}$ |
| Saratov | $10^{-6}$ | $4*10^{-6}$ |
| Kaliningrad - Yuzhny | $10^{-6}$ | $5.3*10^{-6}$ |
| Rostov-Glavny | $4*10^{-6}$ | $1.6*10^{-5}$ |
| Chelyabinsk | $10^{-6}$ | $4*10^{-6}$ |
| Yaroslavl | $10^{-4}$ | $5.4*10^{-4}$ |
| Krasnoyarsk | $5.5*10^{-7}$ | $2.2*10^{-6}$ |
| Novosibirsky | $1.1*10^{-4}$ | $5.5*10^{-4}$ |
| Kievsky | $1.1*10^{-4}$ | $4.4*10^{-4}$ |
| Moskovsky | $5*10^{-6}$ | $2.5*10^{-5}$ |
| Leningradsky | $5*10^{-6}$ | $2.5*10^{-5}$ |
| Kazansky | $10^{-6}$ | $5.2*10^{-6}$ |

**Fig. 6.13** Risks of fire in station buildings located in Russian cities

fire in station buildings located in cities of the Russian Federation (cities of federal importance).

As shown in Fig. 6.13, the most satisfactory situation with station building fire safety is in Samara, Chelyabinsk, Saratov, etc. However, as test calculations show, the levels of risks of station buildings fire and collective risks in Yaroslavl,

Novosibirsk, and Moscow (Kievsky station) are by two orders greater than levels of said risks in "safe" cities. This circumstance requires special attention in order to ensuring the fire safety of these station buildings.

The above examples testify to the great practical importance of the fire risk management methodology.

## 6.7 Occupational Risk Management

Occupational risks are managed in accordance with general risk management scheme set out in item 6.2. Occupational risk management at the Russian Railways is performed at the following levels:

- Production unit: line level (structural divisions), regional level, and central level
- Corporate unit: corporate management center and regional corporate management center.

The occupational risk assessment includes the calculation of occupational risks for the structural division, regional directorate, central directorate, and the Russian Railways as a whole.

Occupational risks are assessed on the basis of:

- Statistics of causes and the number of injuries (for divisions where injuries occurred for 10 years)
- Indirect assessment method (for divisions where there were no injuries), for example, by an expert assessment of the hazard of conditions using questionnaires. It is acceptable to assess occupational risks on the basis of a combination of two parameters: quantitative and indirect (expert) assessments.

Occupational risk assessment includes the following stages:

- Analysis of statistics of injuries in the structural division and quantitative assessment of working environment
- Analysis of the system and methods for ensuring safe working environment based on expert assessments and calculation of the expected number of injuries in a division where there are no injuries
- Assessment of occupational risks in the structural division of the regional directorate
- Assessment of occupational risks in the regional directorate
- Assessment of occupational risks in the central directorate
- Recommendations on risk treatment.

The assessment for each occupation is carried out in three stages:

*Stage 1*—risk assessment in the structural division as a whole. At this stage, the coordinates of risks of the occurrence of minor, severe, fatal, group injuries are sequentially applied to the risk matrix. Next, the level of occupational risk is

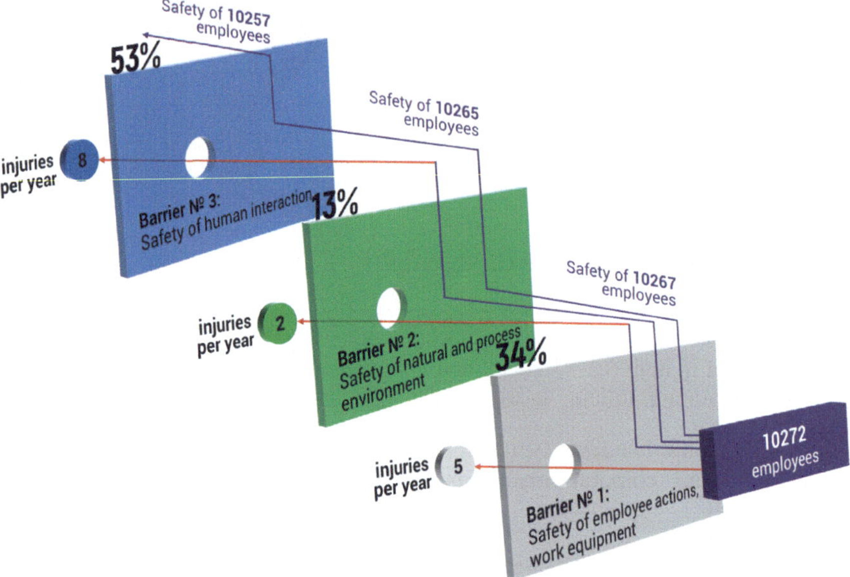


Fig. 6.14 An example of occupational safety barrier disruptions

determined depending on the coordinate. The result of the risk analysis is a risk matrix with marked coordinates (when forecasting for 1 year).

*Stage 2*—assessment of occupational risk by type of incidents. At this stage, the coordinates of risks of injury occurrence for various incidents are sequentially applied to the risk matrix. The risk coordinate is applied to the matrix in the same way as at Stage 1.

*Stage 3*—detailing the occupational risk by sources of injury. At this stage, the coordinates of the risks of injury occurrence for various incidents and from various sources are sequentially applied to the risk matrix.

An indirect (expert) assessment of occupational risks is performed on the basis of a questionnaire survey for safety barriers N 1, 2, 3 and for the layers "organizational measures," "technical measures," and "technical means." The employee safety barrier means a set of measures and technical means that ensure safe working environment which characteristics are specified. In Fig. 6.14, violations of the occupational safety barrier for the profession of "contact network electrician" are shown on the basis of 10-year statistics. Out of the total number of 15 employee's injury, 34% are caused by defective protective gear, 13% are caused by poor technical and technological measures to ensure working environment safety, and 53% (more than half of all injuries) are caused by poor labor management.

Figure 6.15 shows the scale of risks for employees of catenary department of electrification unit.

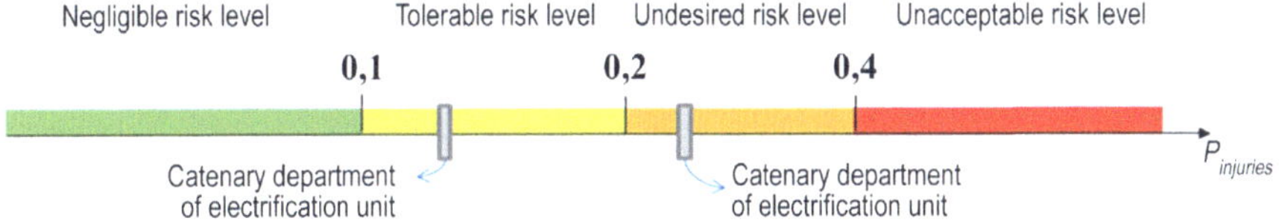


**Fig. 6.15** Scale of risks for employees of catenary system department

Based on the analysis of the state of safety barriers, the number of hazard points is calculated with a subsequent assessment of the expected number of injuries. The occupational risk at the regional level is assessed on the basis of the occupational risk assessments of structural units. The result of the occupational risk assessment at the regional level is a list of assessments of occupational risk in structural divisions for each occupation in the regional directorate.

The risk-based occupational safety management model includes the following seven steps.

*Step 1.* Network risk assessment based on event statistics. Definition of the list of traumatic professions.
*Step 2.* Selecting the directorates where the actual risk assessment will be carried out
*Step 3.* Audit of the management system and processes implemented in a department
*Step 4.* Individual assessment of the hazard of identified states and risks using score cards
*Step 5.* Risk compensation measures
*Step 6.* Reconsideration of individual assessments of employees' occupational risks. Safety reporting
*Step 7.* Recalculation of occupational risks at system (regional) level taking into account individual assessments

The acceptable level of occupational risk is determined after the implementation of risk assessment procedures in accordance with GOST 33433 [11]. The assessment of the level of occupational risk R is correlated with the value $R_{\text{acceptable}}$ (assessment of the acceptable risk). The acceptable level of occupational risk is determined by the criteria of acceptable risk, if $R > R_{\text{acceptable}}$, then such an occupational risk is considered unacceptable.

Implementation of compensatory measures to block hazardous factors includes education of employees on the requirements and workplace safety rules, providing them with tools and protective gear, and increasing the track possession time for repair activity. Actions to reduce the level of occupational risks are formed depending on the level of risk:

- In case of falling into the "undesired" risk area, the following actions can be planned:
  - Extra employee briefing
  - Extra employee training
  - Extra technical inspection, maintenance, etc.

- In case of falling into the "unacceptable" risk area, the following measures should be planned:
  - Modernization or introduction of new technical means
  - Changes in the technological process, etc.

Preference should be given to measures that will ensure such a decrease in the level of occupational risk, that the severity of the negative consequences of injuries will be reduced. As a result, the number of cases with a fatal outcome and serious injuries in the structural unit will decrease.

## References

1. Fasmer, M.: Etimologicheskij slovar' russkogo yazyka (Etymological dictionary of the Russian language. Translation from German, 2nd edn. Stereotype (with additions). Progress, Moscow (1986)
2. Madera, A.G.: Riski i shansy: neopredelennost', prognozirovanie i ocenka (Risks and chances: Uncertainty, forecasting and estimation). URSS, Moscow (2014). isbn:978-5-396-00952-3
3. IEC 62278:2002 Railway applications – Specification and demonstration of reliability, availability, maintainability and safety (RAMS)
4. GOST 33433-2015 Mezhgosudarstvennyj standart "Bezopasnost' funkcional'naya. Upravlenie riskami na zheleznodorozhnom transporte" (Interstate standard "Functional safety. Risk management in railway transportation")
5. Gapanovich, V.A., Shubinsky, I.B., Zamyshlyaev, A.M.: Construction and use of risk matrices in risk management system on railway transport. Dependability Journal. **4**, 56–68 (2011)
6. Cox, L.A. Jr., Huber, W.: Optimal design of qualitative risk matrices to classify binary quantitative risks [abstract]. Proceedings of the Annual Meeting of the Society of Risk Analysis, Boston, 7–10 December, 2008
7. Novozhilov, E.O.: Guidelines for construction of a risk matrix. Dependability Journal. **3**, 73–86 (2015)
8. Pickering, A., Cowley, S.P.: Risk matrices: implied accuracy and false assumptions. Journal of Health and Safety Research and Practice. 2(1) (2010)
9. Gapanovich, V.A., Shubinsky, I.B., Zamyshlyaev, A.M.: Method for assessing risk of the system with diverse elements. Dependability Journal. **16**(2), 49–53 (2016)
10. STO RZD 15.016-2017 Tyagovyj podvizhnoj sostav. Pravila ocenki i upravleniya pozharnymi riskami (Traction rolling stock. Rules for assessing and managing fire risks)
11. STO RZD 02.037–2011 Upravlenie resursami na etapah zhiznennogo cikla, riskami i analizom nadezhnosti (URRAN). Upravlenie stoimost'yu zhiznennogo cikla sistem, ustrojstv i oborudovaniya hozyajstv OAO "RZD" (Management of resources at life cycle stages, risks, dependability analysis (URRAN). Management of the life cycle cost of systems, devices and equipment of the facilities of JSC "Russian Railways")

# Chapter 7
# Resource Management of Railway Transport Facilities

## 7.1 Stages of Life Cycle of Railway Transport Facilities

The life cycle of a railway transport facility is a set of interrelated, consistently implemented processes of setting requirements, creating, using, and disposing of railway engineering that take place over a period of time that starts from the stage of development of the concept of railway engineering and terminates after the stage of its disposal [1, 2, 3, etc.]. The entire life cycle can be divided into three components: (1) Tender stages; (2) Development stages; (3) Stages of operation (Fig. 7.1).

The first five stages (**tender stages**) include the development of *a concept* of the railway transport (RT) facility, an *assessment of its characteristics and operating conditions, risk analysis, development of technical design specifications, and the apportionment of facility requirements.*

***The concept*** of RT facilities may include:

- The results of marketing research
- General requirements for facilities (regulatory requirements and customer requirements)
- Information on the possibilities of production, maintenance and repair or service maintenance
- Feasibility study
- Information on research and development projects (including innovations)
- The results of patent research.

At the stage "***characteristics of the RT facility and the operating conditions,***" we determine whether it is possible and reasonable to develop new railway engineering; we also determine the main characteristics of new RT facilities. In such a case, we use the information on the development of scientific and technical progress, the needs of a customer and industry, market situation, forecasts of changes in the



I. B. Shubinsky, A. M. Zamyshlaev, *Technical Asset Management for Railway Transport*, International Series in Operations Research & Management Science 322,
https://doi.org/10.1007/978-3-030-90029-8_7

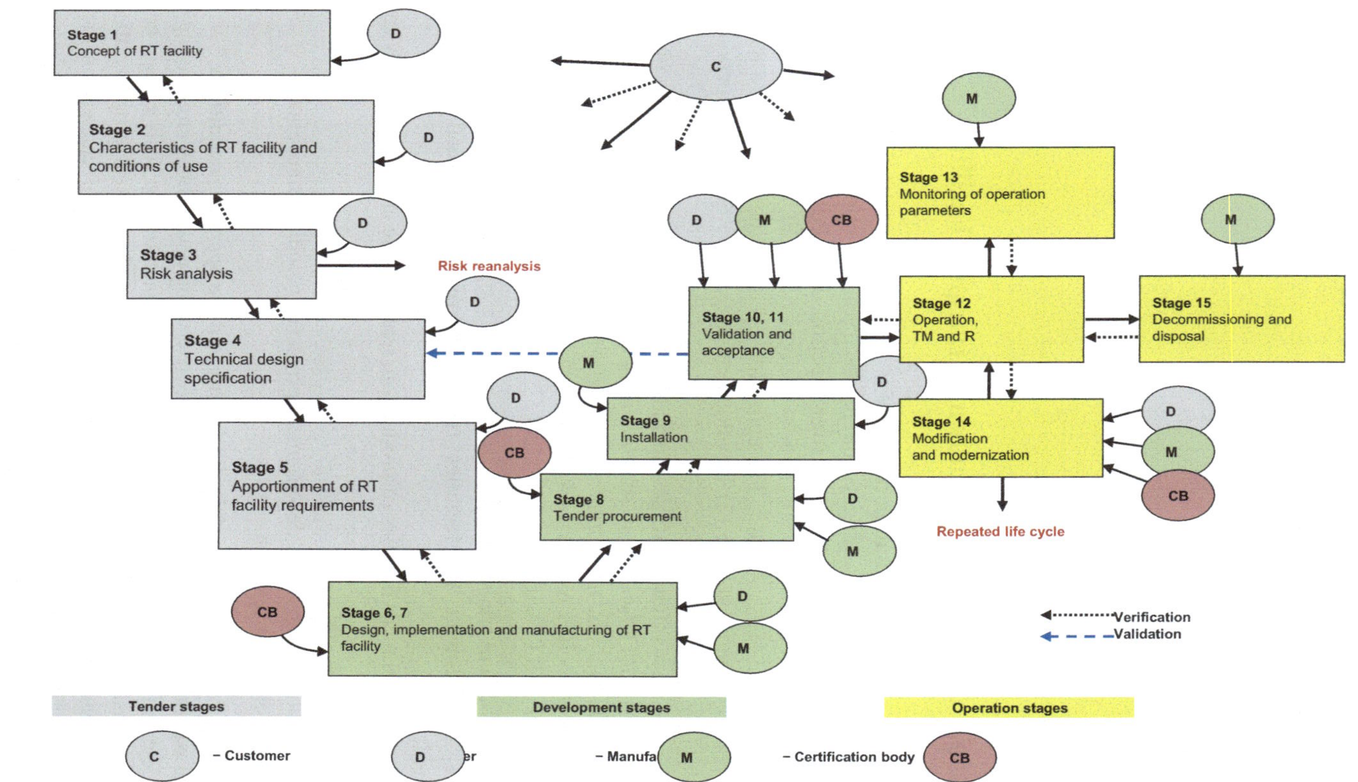


Fig. 7.1 Stages of the life cycle of a RT facility

volume of the transportation process, the results of relevant marketing research in the field of transport services, trends in the development of production processes in a system of maintenance and repair, regulatory documents and legal acts.

***Risk analysis and assessment*** is performed in accordance with the general requirements of GOST 33433-2015 [4].

When identifying hazards and risks we consider the hazards associated with:

- Operation of RT facilities under normal conditions
- Failures of railway engineering or its components
- Operation of railway engineering in an emergency
- Possible misuse of RT facilities
- Interaction of new RT facilities with already operated facilities on railway transport
- Functional features of facilities, including threats to cyber security, hazards related to the functioning of software
- Decommissioning and disposal of RT facilities
- The influence of the human factor
- Occupational safety.

Also, at this stage of the life cycle and in addition to technical aspects, we may consider commercial and financial risks associated with the implementation of RT facility development project.

Risk mitigation actions may include adjusting the technical requirements for facilities or changing measures related to ensuring their dependability and safety. When documenting the risk assessment process, we maintain a hazard record book as a basis for ongoing risk management and make out a report on a risk analysis/assessment. The risk analysis can be repeated at any subsequent stage of the life cycle.

The purpose of the stage "***technical design specifications***" (TDS) is to formulate the entire set of requirements for RT facilities which must be fulfilled during the development (designing). Acceptance of facilities will be carried out by checking them for compliance with these requirements, including mandatory safety requirements. After the approval of the TDS, at the request of the customer, we should prepare a dependability assurance program (DAP) which establishes all the necessary organizational and technical measures, methods, and means to fulfill these specified dependability requirements at the stages of the life cycle. At this stage, we developed a draft DAP in respect to the stages of the life cycle which relate to development. We should also set functional safety requirements for control systems and (or) systems ensuring the safety of the transportation process. These requirements are a part of the technical requirements but can be formalized as a separate document "Functional safety requirements specification." After the approval of the TDS, a safety assurance program (SAP) for these systems should also be prepared.

At the stage of ***apportionment of requirements among the facility and its constituent parts***, the input information is the results of the previous stages of the life cycle. The tasks of this stage are: detailing the requirements of TDS for constituent parts of facilities; development or use of an existing "digital twin" of facility; studying characteristics of the facility under development using a "digital twin"; adjusting the requirements of TDS, risk analysis documents, facility dependability, and safety programs. Here, the concept of "digital twin" means a product similar to the facility under development and includes: (1) the facility model (mathematical, simulation, hybrid or their modifications); (2) software implementation of this model on the chosen platform; (3) the results of verification of this model (preferably the results of its certification); (4) operational manuals for this product.

The tender stages are jointly performed by the customer and the designer of the RT facility.

**The development stages** include the *design and implementation* stage, the *production* stage, the *tender procurement* stage, the *installation* stage of the facility, the *validation* stage, and the *acceptance* stage.

The ***design and implementation*** tasks are as follows:

- Development of preliminary design and engineering design
- Development of working design documentation, technological documentation, operational manuals, repair manual (draft version) for constituent parts, and the Railway Transport (RT) facility as a whole
- Development of software (SW) for functional components and development of program documentation
- Creation of facility engineering prototypes and its testing
- Development of the final version of DAP and SAP and preparing safety evidence for control systems and (or) safety systems.

This stage of the life cycle is jointly performed by the customer, designer, and manufacturer. At this stage, certification of production is carried out by a certification body.

The ***production*** stage includes preparation of the production, mastering of the production, and series production of RT facility. The tasks of the production preparation include: development (refinement) of a set of technological documentation for facility production; development and manufacturing production tooling; provision of infrastructure and providing production environment support; provision of qualified personnel. The tasks of *mastering of production include: manufacture* of preproduction batch; qualification testing of preproduction batch (with supervised operation if required); conformity assessment. *Series production* of RT facilities is a manufactured product: railway engineering or its constituent parts including all the necessary technical documentation and documents confirming compliance that are ready for delivery. This stage of the life cycle is jointly performed by the customer, designer, and manufacturer. At this stage, a certification body certifies production.

The ***tender procurement*** stage is performed by customer companies both at the level of the Russian Railways and at the level of its subsidiaries. Tender procurement is carried out through tenders. Customer companies develop tender documentation taking into account requirements for product dependability and safety prepared on the basis of the URRAN methodology. They announce a tender for the provision of railway facilities or their constituent parts and conduct a tender, where several organizations compete for the opportunity to cooperate with the customer company on certain conditions determined by customer company. This makes it possible to significantly eliminate the corruption component, which often took place when purchasing products. This stage, as well as the production stage is jointly performed by the customer, designer, and manufacturer. A testing laboratory which is entrusted with certification testing of the RT facility is chosen by tender.

At the stage of ***installation*** of facilities, we assemble, install, adjust technical means, construct, perform installation and commissioning works on site which are necessary to create a completed RT facility Also we provide training of operating personnel, and provide the facility with spare parts and tools. It is advisable to separate out this stage of the life cycle only for infrastructure facilities of railway transport, which can be ready for operation only after assembly and adjustment of the entire set of components at a specific place of operation, and not for separate technical means supplied in the form of finished products by manufacturer.

EXAMPLE: *Such facilities, for example, include hump automation systems at marshaling yards.*

The ***validation and acceptance*** stages complete the facility development process. At the validation stage, with the help of the certification body and the testing laboratory, we prove compliance of the facility parameters with the requirements of technical regulations and standards containing mandatory safety requirements applied to railway equipment as well as with TDS, Government decrees, orders of the Russian Federal Service for Technical and Export Control. At the stage of acceptance (confirmation of conformity performed through acceptance), we develop as-built documentation, and if necessary, correct the proof of safety (PS). The RT facility is registered. The facility is commissioned (trial or continuous operation). Both of these stages are jointly performed by the customer, designer, manufacturer, and certification body.

At the stage of ***operation***, we use the RT facility for its intended purpose, provide it with material and technical support, prepare working versions of DAP and SAP for operation, maintain and repair a facility. Also we perform control checks and tests, accept a facility after repair, and assess risks monthly. At the stage of operation, we solve the issues of extending the assigned service life of a facility, the issues of assessing the activities of a division taking into account the safety and dependability of the facility.

At the stage of ***performance monitoring***, we collect, analyze, and evaluate statistical data on performance indicators, collect, analyze, and evaluate statistical data on dependability and safety, monitor hazards and threats, update a list of safety and threats, monitor costs and profits related to railway facility operation, analyze the cost of the facility life cycle, adjust the maintenance and repair system and (or) facility material and technical support. This stage is performed jointly by the customer (operating organization) and the manufacturer.

At the stage of ***modification and modernization,*** we solve the same issues that we perform at the first eight stages of the facility life cycle: we analyze the reasons for making changes and the impact of the change on dependability and safety including the impact on the cost of the life cycle, develop technical documentation for the modernized facility, manufacture a RT facility that has undergone modification (modernization), developed SAP for the stage "Modification and modernization," and confirm the compliance of the modernized facility with the specified requirements. After this stage, the facility can repeat the life cycle.

At the stage of ***decommissioning and disposal***, we plan and then decommission of the facility as well as deregister it. Then the components are de-installed and disposed. Also we developed SAP for control and safety systems at this stage.

## 7.2 Basic Concepts of Facility Physical Resource

There are different definitions of the term "resource" of a facility. Resource is a reserve, a source of something used if necessary, a resource is a means, an opportunity to implement something, a resource is the possible operating life of a machine's indicated in its technical passport, an Internet resource, etc. There are economic, financial, material, labor, environmental, information resources, computing, network resources, physical resources of technical systems (its dependability resource), etc.

All types of resources are divided into two groups: renewable and non-renewable resources. For example, a group of non-renewable resources includes resources, the reserves of which may deplete in the near future if they are used with the current rate. On the other hand, renewable resources are natural resources whose reserves either recover faster than they are used, or do not depend on whether they are used or not. The dependability budget of a railway transport facility with a reserve refers to a category of a renewable resource of a technical facility (system). The dependability budget of a railway facility without redundancy refers to a category of a non-renewable resource. We are talking about such facilities of railway transport as rails, fastening sleepers, points, catenary system lines, transmission line pillars, and many other facilities. Their dependability influenced by physicochemical processes occurring in the materials of the constituent parts changes over time. This is due to the fact that the material properties depend on a large number of factors (operating mode, humidity, temperature, etc.). Table 7.1 shows the mechanism of various types of factors affecting the dependability of elements of railway transport facilities.

Typically, technical equipment (element) is affected by factors of several groups at the same time. The mechanism of various types of factors affecting the dependability of elements still cannot be described in any detail. This is one of the reasons why the way of statistical description of the processes affecting the system dependability is not only necessary, but also the only possible one [5].

Professor N.M. Sedyakin proposed the principle (law) of dependability according to which the notion of a dependability resource of an element (system) is set in the form: $r(t) = \int_0^t \lambda(z)dz$, where $\lambda(t)$ is the failure rate of the element. The law itself is formulated as follows: "The dependability of an element (system) depends on the length of its service life spent in the past and does not depend on the way it was spent" [6]. N.M. Sedyakin's principle is still the only powerful tool for assessing the dependability of elements and systems operating under variable load conditions.

**Table 7.1** Characteristics of the main types of loads of railway engineering

| No. | Load type | Characteristic | External demonstration |
|---|---|---|---|
| 1 | 2 | 3 | 4 |
| 1 | Strength loading | The impact of an external concentrated and distributed force, which causes a corresponding response in a material in the form of internal forces distributed over the section | Deformation |
| 2 | Thermo-mechanical loading | Thermal expansion or compression of a material during heating or cooling when there is no space for such a change in geometric dimensions | Thermal elongation or compression when deformation is limited |
| 3 | Frictional impact | Detachment and transfer of material particles under the influence of friction forces | Change in the size and/or shape of mating parts as a result of wear |
| 4 | Corrosion and erosion impact | A gradual change in the dimensions of a technical equipment (element) or in material properties as a result of physicochemical interactions of a technical equipment (element) with an external aggressive environment | Changes in the size and/or shape of mating parts as a result of corrosion |
| 5 | Electro-magnetic impact | Changes in the physical and chemical properties of materials under the influence of current | Current flowing or presence of external electromagnetic radiation |

This principle (law) served as the basis of the URRAN system for assessing the residual life and making a decision on extending the service life of facility, since it make it possible to avoid labor-intensive activity when assessing the current resource. Also it often helps to avoid performing impossible tasks to determine the way facility life was spent in the past.

## 7.3 Methodological Bases for Extending the Service Life of a Technical Facility of Railway Transport

### *7.3.1 General Provisions*

Upon expiration of the service life specified in the technical documentation, we assess the possibility of further operation of a facility by assessing its technical condition and resource. The basic approach adopted for the URRAN system is based on the principle of "safe operation according to technical condition." According to this principle the assessment is performed by parameters of the facility technical condition that ensure dependable and safe operation of the facility in accordance with the regulatory and (or) design (project) documentation. The residual life (in accordance with said principle) is assessed by determining parameter of the technical condition. The decision to extend the residual life is made according to the criteria "safety," "dependability," and "efficiency." The criterion "safety" includes indicators which values are equal to or exceed the maximum acceptable indicators. The criterion "dependability" includes indicators which values do not exceed the maximum acceptable values. The criterion "efficiency" includes economic indicators, which determine whether it is rational to continue the operation of an element of the facility or replace it, provided that the assigned service life of the facility element has been expired but the actual residual life (determined on the basis of the criteria "safety" and "dependability") is more than 5 years.

This paragraph discusses the task of extending a non-renewable resource of dependability of the facility (element and system). This task is especially relevant for track facilities and a significant number of facilities of the railway electrification and power supply complex (Transenergo facilities). The task of extending the physical resource of facilities of signalling and remote control systems and communication complex is of secondary importance for the following two main reasons: (1) these facilities are highly redundant, therefore, their dependability resource is renewable; (2) the vast majority of the elements of these facilities are electronic means (more than 80%), which failure rate does not change over time, its failures are sudden and there is no element wear and aging. Therefore, as for facilities of signalling and remote control systems and communication complex, it is relevant to consider a functional resource (see item 7.5).

The general algorithm for assessing the technical condition of an infrastructure facility provides for the sequential implementation of the following stages:

- Collection and analysis of the initial technical information about the facility
- Operational (functional) diagnostics
- Expert inspection of facility technical condition
- Analysis of damages, establishment of their mechanism and control parameters of facility technical condition
- Establishment of principles of changes in the control parameters of the technical condition, limit states and their criteria
- Analysis of failures and limit states, assessment of the consequences and criticality of failures
- Processing of the received data and resource forecasting
- Justification of options for decisions on the possibility of further facility operation.

As parameters of the technical condition, we take parameters the change of which (individually or in some aggregate) can bring the facility into an inoperable or limit state. We assess the technical condition parameters and choose the control parameters based on the results of the analysis of design documentation (drawings and diagrams), technical documentation (technical passport and instructions), data from diagnostic track machines such as Integral, Tvema, etc., the diagnostic car of the VIKS and VETL laboratories, equipment for permanent technical diagnostics, as well as on results of rounds with inspections. The residual life is assessed on the basis of the established patterns of change in control parameters, formation of defects

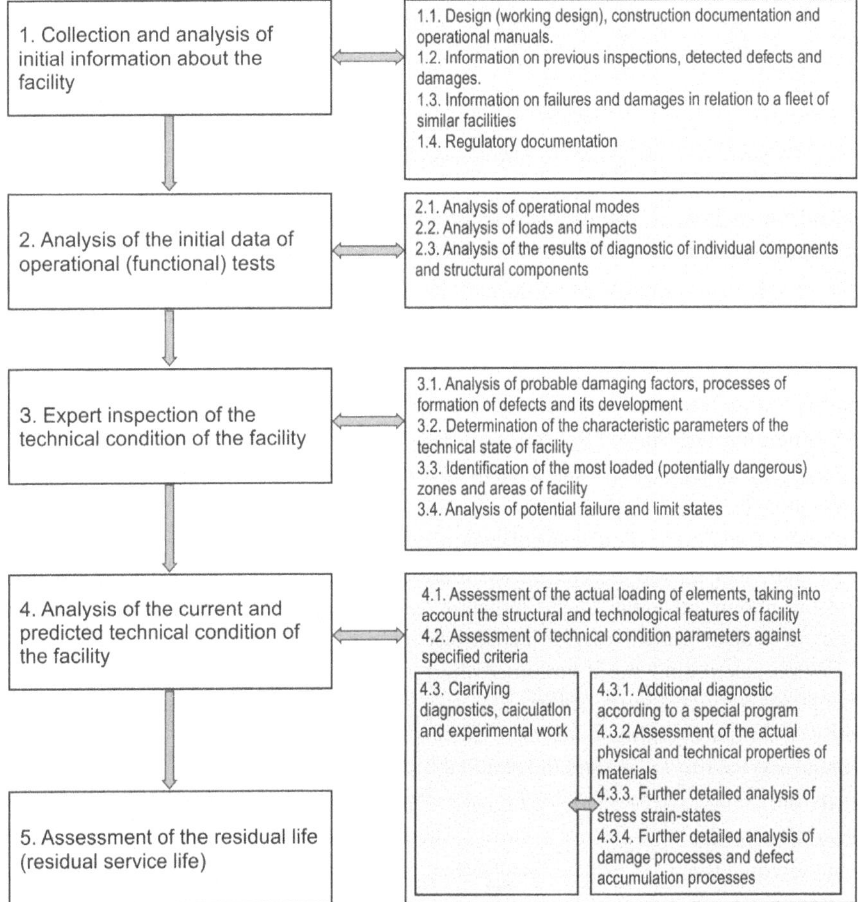


**Fig. 7.2** General algorithm for assessing the residual life

and its development and (or) on the basis of the results of functional indicators measurement.

The general algorithm for assessing the residual life of railway transport facilities is shown in Fig. 7.2.

The residual life of the facility is determined based on the analysis of the operating conditions, the results of technical diagnostics, and the criteria of a limit state. When residual life is estimated based on consideration of several criteria of the limit state, then we assign residual life according to the criterion that determines the minimum residual life. *The calculation* of *residual life* is made for facilities:

- The actual service life of which exceeds the assigned service life
- The actual service life of which is less than the assigned service life by 5 years
- Defective and extremely defective facilities.

### 7.3.2 Control and Evaluation Maps

Residual life of the facility $T_{res}$, is often determined by control and evaluation maps (AEM) for its elements. During the operation of elements, it is necessary to periodically inspect, test, measure, monitor, check, and compare the obtained results of parameter measurement ($S_{factual}$) with its acceptable value ($S_{acceptable}$). This principle is laid down in the design and construction of the AEM. Assessment charts are developed on the basis of the above criteria "safety," "dependability," and "efficiency."

For example, if the factual assigned service life of facility element has been expired, but the residual life of the element allows its operation for another 10 years, then it is advisable to assess the economic efficiency of the further operation of the element of the facility or its replacement. In the process of filling out the AEM, we determine the indicators "$B_i$" (the number of points awarded to the parameter). If $S_{factual} \geq S_{acceptable}$, then we assign one hundred points to the parameter "$B_i$" (indicator $B_i = 100$). If $S_{factual} < S_{acceptable}$, then the indicator $B_i$ is calculated depending on the wear and according to the formula: $B_i = \frac{S_{factual}}{S_{acceptablen}} 100$. Then all the parameters for the selected element are calculated and recorded into the AEM, and the maximum value of the indicator "$B_i$" is selected. The residual life of the element is determined by the selected maximum value of the indicator "$B_i$".

When calculating the duration of the residual time life, it is assumed that the acceptable service life of facility $T_{acceptable}$ (established for reasons of its impossibility of further operation in terms of unsatisfactory safety) is equal to the sum of the factual service life $T_{factual}$ and the residual life $T_{res}$. Hence, $T_{res} = T_{acceptable} - T_{factual}$. In its turn, $T_{acceptable} = \frac{S_{acceptable}}{V}$, $T_{factual} = \frac{S_{factual}}{V}$, $V = \frac{S_{factual}}{T_{factual}}$, where $V$ is the defect development rate.

Consequently, $T_{acceptable} = \frac{S_{acceptable}\,T_{factual}}{S_{factual}}$.

Then

$$T_{res} = T_{acceptable} - T_{factual} = \frac{S_{acceptable} T_{factual}}{S_{factual}} - T_{factual} = T_{factual}\left(\frac{S_{acceptable}}{S_{factual}} - 1\right)$$

Calculations are converted into points: $\frac{S_{acceptable}}{S_{factual}} = \frac{1}{B}$ (if $B$ is proportion of point); $\frac{S_{acceptable}}{S_{factual}} = \frac{100}{B}$ (if $B$ is percent).

Then the residual life of the element $T_{res}$ determined by the formula:

$$T_{res} = T_{factual}\left(\frac{100}{B} - 1\right), \quad \text{(if } B \text{ is percent)} \tag{7.1}$$

where $B$ is maximum number of points. $B = B_{i\ max}$

If, in the process of filling in the AEM, the points awarded to the one of the parameters $B_i = 100$, then it is considered that the element of the facility in regard to this parameter has exhausted its entire service life and $T_{res} = 0$. If the residual life is more than 10 years according to the calculation, then the residual life is assumed to be unlimited.

In case a physical facility at the stage of creation was assigned to the calculation of one element, then the residual (full) life of the facility is taken as the residual (full) life of this element of the facility. If more than one calculated element was assigned to the same physical facility at the creation stage and each of these elements corresponds to a certain type of defect or damage accumulation mechanism, then the minimum of the obtained values is taken as the final calculated evaluation of the facility life.

Based on the results of the analysis of the operability and life of the facility, there are three options for decisions on further activities:

1. Continuing the operation of the facility without changing the mode until complete exhausting or achieving the specified indicators of the residual life (or during a specified calendar time within the available residual life)
2. Continuing the operation of the facility, provided that its load is reduced to a level ensuring that the available calculated residual life is not lower than the assigned extended life. Such a solution should be considered as a temporary measure aimed at maintaining the partial operability of the facility for a limited period
3. Termination of operation in order to conduct partial repairs of the facility or replace it.

The cause for continuing the operation of the facility in the previous mode is the excess of the calculated value of the residual life in comparison with the extended life assigned on the basis of the results of analysis.

Residual life of the facility $T_{res}$ is determined by the AEM for all elements. The elements are ranked with a breakdown by residual life: up to 5 years; from 5 to 10 years; more than 10 years. The residual life of a facility consisting of diverse elements is determined taking into account the weight coefficients that take into account the influence of an individual element of the facility on the residual life of the entire facility.

## 7.4 Criteria for Extending the Service Life of a Facility

Extending the service life of a facility is possible if it is not in a limit state. According to GOST 32192 [7], the limit state is the state of railway engineering in which its further operation is unacceptable or impractical in accordance with a risk assessment. The limit state criteria should characterize the unacceptability of the operation of a railway transport facility based on a risk assessment or the impracticality of its operation based on an assessment of economic efficiency.

*The unacceptability* of using a facility for its intended purpose is determined *by non-compliance with the specified safety requirements* for its operation due to exceeding the acceptable level of risk. The reason for this is the deterioration of the technical condition of railway engineering due to the expiration of its life during

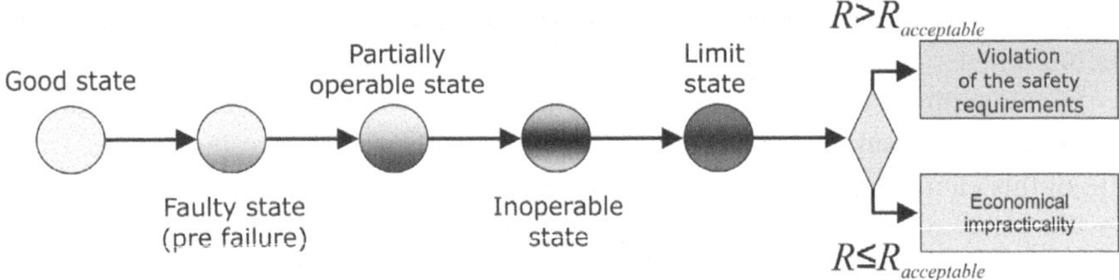


**Fig. 7.3** Scheme of possible transformation of facility states during operation

operation. A change in the technical state of railway engineering affect on its performance characteristics, in particular, on the indicators of dependability and safety of its functioning. The parameters determining the operability of each element (individual component, unit) of technical equipment deteriorate in different ways: some—faster, others—slower, and still others—discretely. Some parameters can deteriorate "autonomously" without affecting other characteristics, the deterioration of others leads to multiple failures. Besides, the longer the service life of a technical equipment the higher the probability of a failure. The technical equipment of railway transport under the influence of the loads (given in Table 7.1) during its operation transfers into a state in which its further operation is either dangerous or economically impractical. Figure 7.3 shows a diagram of a possible transformation of the states of a facility during operation. The transition of a technical equipment from one initial technical state to another state usually occurs as a result of damages or failures. The transition of technical equipment from an operable (serviceable, faulty, pre-failure, and partially operable) state to an inoperable state is called a failure, and the transition to a limit state is called depletion of life. If the risk of a hazardous state of the facility being in limit state R exceeds the acceptable level, then the facility is decommissioned. Otherwise, we assess the economic impracticality (or, conversely, the practicality) of the further operation of the facility.

Let us consider a situation when the risk of a hazardous state of a facility *R* exceeds the acceptable level and its further operation is *unacceptable according to safety requirements*. Figure 7.4 shows that if there is a rational organization (strategy *i*) of the maintenance and repair (TM and R) system, it is possible to maintain the probability of facility failure-free operation at an acceptable level, at which its failures are unlikely or extremely rare. This is possible even after the expiration of the assigned service life. If TM and R system (strategy *j*) is insufficiently effective, the facility may turn out to be at an unacceptable level of safety even within the assigned service life.

When categorizing failures of a technical facility according to the severity of their consequences, we should take into account at least the following factors in various combinations:

- The failure danger (taking into account immediate and long-term consequences) to the life and health of people (including those not directly involved in the operation of the technical facility), the environment, the integrity and safety of the technical facility itself, other property and material assets

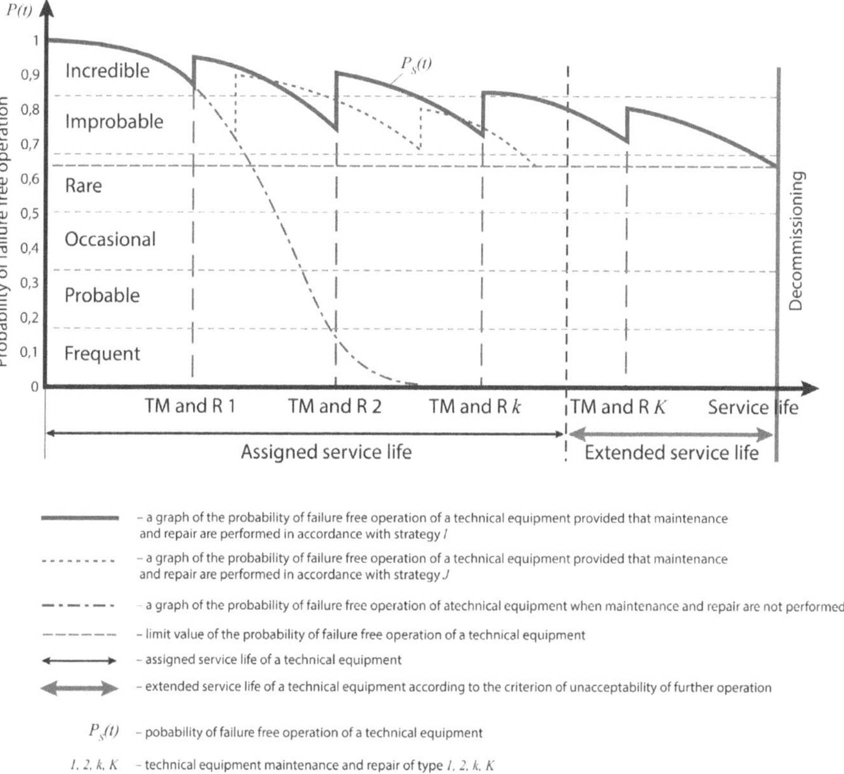


**Fig. 7.4** Dependence of the probability of failure-free operation of a technical equipment on the service life

- The impact of failure on the quality of technical equipment functioning, completeness of performing its assigned functions, possible damage of any kind (material, moral, political, etc.) caused by a decrease in the quality of functioning of the technical equipment or its failure to perform certain functions (assigned tasks)
- The rate of development of adverse consequences of failure, which determines the possibility of taking appropriate protection measures against them.

The size of failure consequences directly affects the amount of measures that should be taken to avoid this failure. In other words, the more serious the possible consequences of a failure, the more efforts should be made to eliminate it. In fact, it is necessary to develop a strategy for application of technical equipment maintenance and repair not to prevent failures by itself, but to avoid or minimize possible consequences. The direct costs play a special role in eliminating any failure. They can be expressed not only in the form of repair costs, but also in the form of fines, penalties, etc.

Based on the analysis of the types and severity of the failures consequences, we determine the most critical technical equipment (individual components, parts, and elements), and choose different action options for each technical equipment. A measure of the severity of the consequences and the probability of failure occurring over a given period of time is the risk assessed for each critical element. Then, for the object as a whole, we determine the integral risk within a given time. The risk associated with the functioning of the facility during operation tends to gradually increase due to the deterioration of its technical condition. Therefore, it should be recalculated periodically.

The economic inexpediency of operation is determined by a decrease in the efficiency of using the facility. It is due to an increase in the cost of its life cycle (hereinafter—LCC) or its functional (moral) obsolescence (see item 7.5).

*The criterion of economic inexpediency of operation* lies in the fact that it is more profitable to replace the operated facility with a new one at some point in time *T*. This is related to the fact that with an increase in the service life, the costs of facility operation and TM and R increase due to the physical wear of the facility. To make a decision in accordance with this criterion, it is necessary to compare the average annual life cycle cost of the operated technical facility $\overline{\mathrm{LCC}}_{\mathrm{op}}$ and new technical facility $\overline{\mathrm{LCC}}_{\mathrm{new}}$.

The obtained values of the average annual $\overline{\mathrm{LCC}}$ of the operated technical equipment for a given period, $\overline{\mathrm{LLC}}_{\mathrm{op}}\left(T_{\mathrm{gp}} - T_j\right), \textit{где}\ \left(0 \leq T_j < T_{\mathrm{gp}}\right)$ are compared with the minimum value of average annual $\overline{\mathrm{LCC}}$ for the new technical facility, $\mathrm{LCC}_{\mathrm{new}}(T_{\min})$. If $\overline{\mathrm{LLC}}_{\mathrm{op}}\left(T_{\mathrm{gp}} - T_j\right).> \mathrm{LCC}_{\mathrm{new}}(T_{\min})$, then it means that it is reasonable to replace the operated technical facility with a new one at the moment of time $T_j$. In the area $\overline{\mathrm{LLC}}_{\mathrm{op}}\left(T_{\mathrm{gp}} - T_j\right) \leq \overline{\mathrm{LCC}}_{\mathrm{new}}(T_{\min})$ it is economically reasonable to extend the service life of the operated technical facility.

This criterion is shown graphically in Fig. 7.5, where function of economic efficiency is as follows: $F\left(T_j\right) = \overline{\mathrm{LCC}}_{\mathrm{new}}(T_{\min}) - \overline{\mathrm{LLC}}_{\mathrm{op}}\left(T_{\mathrm{gp}} - T_j\right)$.

Formulas for determining the average life cycle cost of the operated and new facilities are given in [8].

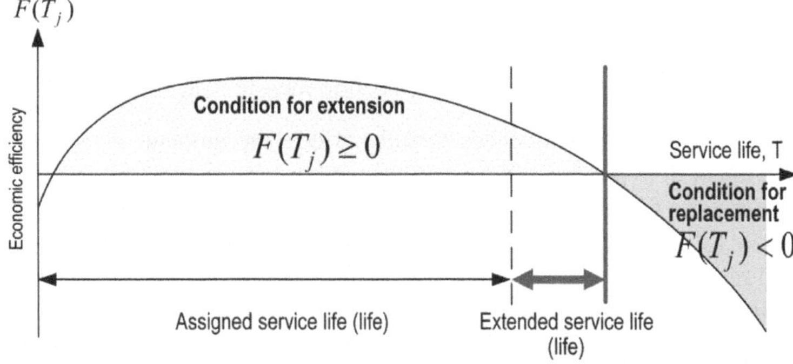


**Fig. 7.5** Function of the effectiveness of extending the service life of the operated technical facility

## 7.5 Functional Resource

Most of the failures in the railway signalling and remote control facilities (SRCF) complex as well as in the railway communication complex are of a sudden nature and do not depend on the deterioration and wear and tear of technical equipment. As for the technical facilities of these complexes of the Russian Railways, when making a decision on prolongation or termination of their operation, it is important that their functional capabilities comply with the current requirements for train-handling and carrying capacity of railways. This refers to a functional resource.

*A functional resource* is understood as a resource of an SRCF or communication facility in terms of the quantity and quality of functions implemented. This resource determines the possibility of using said facility in various operating conditions [9]. The functional resource is assessed on the basis of a set of quantitative and qualitative indicators that take into account the specifics of the functional requirements for railway SRCF or communication facilities at a station or in open line. The quantitative requirements include: the train-handling and carrying capacity of a railway line, the dependability and safety of the facility. The quality requirements include mandatory and additional functional requirements for SRCF (communication) systems. When assessing the functional resource of the facility, the following indicators are used: the utilization coefficient of the train-handling capacity of a railway line, the failure rate and dangerous failure rate of the facility. These indicators are separately calculated or set for a new facility in operation.

Mandatory quality indicators are formed on the basis of functional requirements for facilities at the station and in open line, taking into account the class and category of railway line. Additional qualitative indicators serve to assess the facility functional capabilities for expansion of its functional requirements specified as well as technical and technological capabilities.

The utilization coefficient of the train-handling capacity of a railway section equipped with SRCF is calculated by the following formula. This formula is applicable to both new SRCF and those in operation:

$$k = \frac{n_{\mathrm{av}}}{n_{\mathrm{req}}} \tag{7.2}$$

where:

$n_{\mathrm{av}}$ is the available train-handling capacity
$n_{\mathrm{req}}$ is the required train-handling capacity.

Available train-handling capacity of new railway SRCF is determined with respect to the expected operating conditions. Available train-handling capacity of railway SRCF in operation is determined with respect to the factual operating conditions.

We determinate the design value of the failure rate only for new SRCF. As for SRCF systems in operation, we assess their residual life. It is assessed on the methodological basis set out in item 7.3.1 and in item 7.3.2.

**Table 7.2** General characteristics of SRCF and conditions of its operation

| Characteristic | Note |
|---|---|
| System type | A type |
| Climatic zone | Zone number is from 1 to 9 |
| Class and category of railway line | Class: 1, 2, 3, 4, 5. Category: HSR, passenger line, freight line, heavy-traffic line, heavy haul rail, low-density line |

**Table 7.3** The resulting indicators used for assessment of the functional resource of systems

| Resulting indicator | SRCF system | |
|---|---|---|
| | New SRCF system | SRCF system in operation |
| $Z_1$ | $Z_1^N = \begin{cases} k_C \geq 1.15 \Rightarrow k_C \\ k_C < 1.15 \Rightarrow 0 \end{cases}$ | $Z_1^{\text{op}} = \begin{cases} k_f \geq 1 \Rightarrow k_f \\ k_C < 1 \Rightarrow 0 \end{cases}$ |
| $Z_2$ | $Z_2^N = \begin{cases} \lambda_C \leq \lambda_a \Rightarrow \frac{\lambda_a}{\lambda_C} \\ \lambda_C > \lambda_a \Rightarrow 0 \end{cases}$ | This indicator is not calculated for the SRCF systems in operation |
| $Z_3$ | $Z_3^N \Rightarrow 1$, if there is a document of a standard form<br>$Z_3^N \Rightarrow 0$, if there is no document of a standard form | $Z_3^{op} = \begin{cases} \lambda_f^0 \leq \lambda_a^0 \Rightarrow \frac{\lambda_a^0}{\lambda_f^0} \\ \lambda_f^0 > \lambda_a^0 \Rightarrow 0 \end{cases}$ |

We use the design value $\lambda_d$ of failure rate of the SRCF system as the initial data for the calculation. This design value is determined by the manufacturer and is set as a technical and operational indicator of the new SRCF system. Also we use the information on expected operating conditions of (put into a Table similar to Table 7.2) as the initial data.

The design value of the failure rate of the SRCF system under the expected operating conditions is calculated using the formula:

$$\lambda_c = \lambda_d k_{cl} k_{load} \tag{7.3}$$

where

$k_{Cl}$ is the climatic correction coefficient

$k_{load}$ is the correction coefficient taking into account the load of the railway section.

To calculate the factual value of the failure rate we use failure statistics. The value of the acceptable rate of dangerous failures is determined on the basis of the acceptable values for the SRCF at the station and the open line, taking into account their characteristics. We determine the factual value of the dangerous failures rate only for SRCF systems in operation. As for new systems, we use safety proof procedure.

Table 7.3 presents the resulting quantitative indicators of the SRCF system [10].

The integral quantitative indicator is determined on the basis of the results of calculations in accordance with Table 7.3:

- For new systems—according to the formula:

$$Z = Z_1^N Z_2^N Z_3^N \tag{7.4}$$

- For systems in operation—according to the formula:

$$Z = Z_1^{\mathrm{op}} Z_3^{\mathrm{op}} \tag{7.5}$$

If at least one of the resulting indicators is zero, then the integral indicator is also zero. In other cases, the values of the indicator $Z \geq 1$.

The size of reserve by quantitative criteria is calculated based on the formula:

$$\Delta Z = \begin{cases} Z \geq 1 \Rightarrow Z - 1 \\ Z = 0 \Rightarrow 0 \end{cases} \tag{7.6}$$

When assessing quality indicators, an integral indicator is formed separately for the mandatory and additional quality indicators:

- mandatory quality indicators:

  $Z_m = 1$ if all $Z_i$ are defined as "yes";
  $Z_m = 0$ if at least one $Z_i$ is defined as "no."

- additional quality indicators:

$$Z_{\mathrm{Ad}} = \begin{cases} 0.75 \leq \eta \leq 1 \Rightarrow 3; \\ 0.5 \leq \eta < 0.75 \Rightarrow 2; \\ 0.25 \leq \eta < 0.5 \Rightarrow 1; \\ 0 \leq \eta < 0.25 \Rightarrow 0 \end{cases} \tag{7.7}$$

with $\eta = \frac{m_z}{M_z}$, where:

$m_z$ is the number of additional quality indicators with "yes" value for a railway line of the corresponding class and category

$M_z$ is the total number of additional quality indicators for a railway line of the corresponding class and specialization.

The integral indicator characterizing the compliance with the mandatory functional requirements for the system is calculated by the formula:

$$R = \Delta Z Z_m \tag{7.8}$$

Based on the calculated values of the integral indicators $\underline{R}$ and $Z_{\mathrm{Ad}}$, we assess the functional resource of the SRCF system using the matrix presented in Table 7.4.

The results of assessing the functional resource of the SRCF (communication facilities) in operation, together with the results of assessing their residual life, are

**Table 7.4** Assessment of the level of the SRCF systems functional resource

| | Indicator $Z_{Ad}$ | | | |
|---|---|---|---|---|
| Indicator $R$ | $Z_{Ad} = 3$ | $Z_{Ad} = 2$ | $Z_{Ad} = 1$ | $Z_{Ad} = 0$ |
| $R > 3$ | High | High | High | Medium |
| $0.22 < R \leq 3$ | High | Medium | Medium | **Minor** |
| $0 < R \leq 0.22$ | Medium | **Minor** | **Minor** | **Minor** |
| $R = 0$ | **Not available** | **Not available** | **Not available** | **Not available** |

the basis for making a decision to extend their service life, or making a decision on their modernization or replacement.

As for systems that do not have a functional resource, a number of the following additional recommendations are given:

- If $Z_1 = 0$ and the train-handling capacity of a railway section is unsatisfactory, then some elements of the infrastructure need to be replaced to ensure the required train-handling capacity
- If $Z_2 = 0$, which indicates that the level of system dependability is insufficient, the use of a new system under those circumstances is not recommended
- If $Z_3 = 0$, then the level of system safety is unsatisfactory. It is not recommended to use the new SRCF system under those conditions. Upgrading or replacement of the system in operation is recommended
- If $Z_m = 0$, then the functionality of the SRCF system does not meet the requirements of the standards. In this case, it is recommended to replace the system.

## 7.6 Management of Technical Resources of Infrastructure Facility During Operation

Railway transport is a complex technical system; therefore, the main component of resource management is making decisions on the technical maintenance of railway infrastructure facilities. The resource management of rolling stock is not considered in this book since the vast majority of rolling stock is under maintenance.

Currently, one of the main criteria of the effectiveness of resource management is the assessment of the life cycle cost (LCC) of a facility.

To manage the technical maintenance of railway transport facilities, we should solve the following tasks:

- Estimation of the cost of the facility technical maintenance (TM) and the cost of its repair (R)
- Making a choice of the best strategy for planning work on maintenance and repair in accordance with the criterion of economic efficiency.

The period for analyzing the LCC should include all stages of the system life. However, in terms of rail systems, it is reasonable to consider the duration of the operational stage as the service life when analyzing the life cycle. This is due to the fact that operational stage covers the largest proportion of the system's life cycle.

Since the railway infrastructure and especially the track is an expensive asset with a long service life, it is necessary to guarantee the economic efficiency of a long-term project and decisions on maintenance and repair. Technical maintenance policies and budgetary constraints play a major role when choosing alternative TM and R strategies. They serve as key input data when making a decision on a specific strategy for TM and R. The TM and R strategy that provides the lowest LCC is considered as a cost-effective decision during system operation.

As for track facilities, signalling and remote control facilities, and electrification and power supply facilities, the task of optimizing the LCC reduces to finding such a repair practice which provide the minimal average annual cost of the life cycle of the infrastructure facility (subject to the established restrictions):

$$\begin{cases} S_{\text{total}}^{av} = \dfrac{S_{\text{rec}} + \sum\limits_{i=1}^{n} \left( S_{\text{repairs}}\ (i) + S_{\text{M}}(i) + S_{\text{failures}}(i) \right)}{n} \longrightarrow \min\,; \\ \eta(n) \leq C, \end{cases} \tag{7.10}$$

where:

$S_{\text{total}}^{\text{av}}$ is a part of the life cycle cost, the value to be optimized
$S_{\text{rec}}$ is the cost of reconstruction
$S_{\text{repairs}}\ (i)$ is the cost of repairs in year $i$
$S_{\text{M}}(i)$ is the cost of maintenance in year $i$
$S_{\text{failures}}(i)$ is the cost of failure in year $i$
$n$ is the duration of the life cycle in years
$\eta$ is a failure rate, dependability indicator
$C$ is the required level of dependability.

With an increase in the number n of years of facility operational life, the following costs increase: the average annual direct costs of current maintenance, the average annual cost of unscheduled repairs, and the average annual cost of downtime. The minimum costs are achieved when the amount by which the cost of the listed components increases per year is equal to the amount by which the average annual cost of capital repairs decreases per year. Of the listed components, only the average annual cost of capital repair is known, and, therefore, its derivative can be obtained. The average annual cost of failures and the average annual cost of maintenance may increase or remain constant.

Optimization of the average annual cost of the life cycle of an infrastructure facility at the operational stage is based on structure of costs (shown in Fig. 7.6).

The following options are possible to change each of these two values, provided that the second value is constant:

- Increase in average annual direct costs of facility maintenance with constant average annual cost of failures. The average annual cost of failures is constant

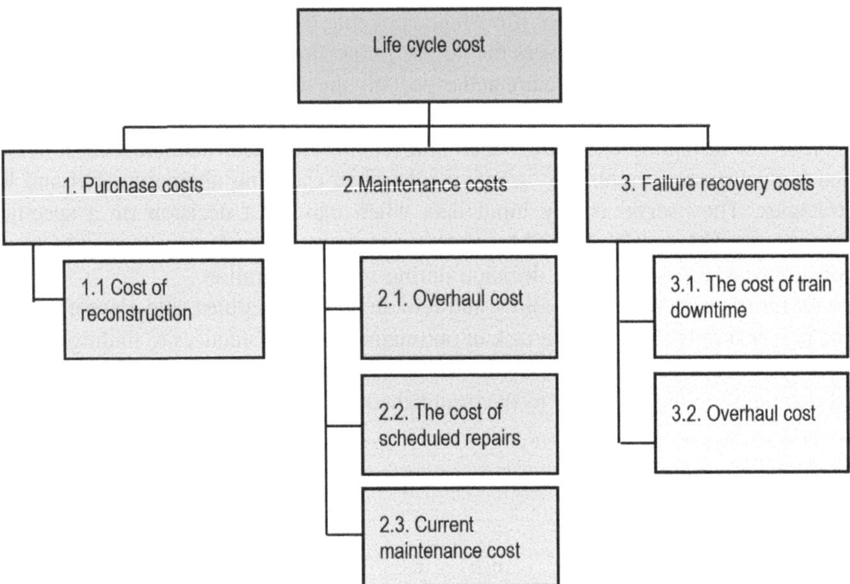


**Fig. 7.6** The structure of the cost of the life cycle of an infrastructure object at the operational stage

when the failure rate does not increase over the years and does not reach the limit values. This can happen if the maintenance of the facility is intensified in order to prevent the increase in failures, thus, the average annual cost of the maintenance increases over the years. Therefore, if the failure rate of a facility is practically constant, then it is cost-effective to schedule overhaul when the cost of maintenance is equal to or close to the sum of the average annual maintenance cost for all previous years and the average annual cost of overhaul.

- Increase in the average annual cost of failures with constant average annual costs of the facility current maintenance. The average annual cost of unscheduled repairs and the average annual cost of downtime depend on the failure rate. When the failure rate increases rapidly it reaches the limit values and, therefore, leads to the need to assign overhaul based on the value of the failure level.
- In general, both the average annual direct costs of the facility current maintenance and the average annual cost of failure can increase.

The approaches applied to the management of infrastructure facilities resources make it possible to increase the efficiency of their current maintenance and overhaul by managing the life cycle cost, provided that the required level of dependability of technical means is ensured.

Compliance with safety and dependability criteria is the main priority in the complexes of signalling and remote control facilities, electrification and power supply facilities. One of the main tools for managing resources in these complexes is the Programs for the reconstruction of their fixed assets of the Russian Railways and Transenergo, respectively.

When making decisions on the inclusion of an activity in the preliminary list of the Program for the reconstruction of fixed assets, we calculate the cost of the facility life cycle. This calculation is performed for two options: for the case of implementation of the activity and for the case of refusal to implement the activity (alternative option). For each project, we calculate alternative implementation options (if any) and evaluate the average annual LCC according to the formula:

$$\overline{\mathrm{LCC}} = \frac{\mathrm{LLC}}{\frac{1}{d}\left(1 - \frac{1}{(1+d)^T}\right)} \tag{7.11}$$

where:

LCC is the cost of the facility life cycle for the entire service life

$T$ is the value of the service life of a new device, years

$d$ is the discount rate.

If there are alternatives for each project, we select the option with the lowest average annual LCC.

The economic effect of each activity (included in the preliminary list of the Program for the reconstruction of fixed assets of the complex of facilities) is determined by the difference between the average annual facility LCC when implementing the alternative option and the average annual facility LCC when implementing the activity included in the Program for the reconstruction of fixed assets of the complex of facility. This economic effect is calculated by the formula:

$$\mathrm{EE}^{i}_{\mathrm{activity}} = \overline{\mathrm{LCC}}^{i}_{\mathrm{al}} - \overline{\mathrm{LCC}}^{i}_{\mathrm{activity}} \tag{7.12}$$

where:

$\mathrm{EE}^{i}_{\mathrm{activity}}$ is the economic effect of the ith activity of the Program for the reconstruction of fixed assets of the complex of facilities

$\overline{\mathrm{LCC}}^{i}_{\mathrm{al}}$ is the average annual cost of the life cycle when implementing the alternative option

$\overline{\mathrm{LCC}}^{i}_{\mathrm{activity}}$ is the average annual cost of the life cycle when the implementing activity included in the Program for the reconstruction of fixed assets of the complex of facilities.

The criterion for the economic efficiency of the overall investment program for complexes of signalling and remote control facilities of JSC Russian Railways and Transenergo (for a given amount of investments) is the achievement of the maximum total economic effect of the implementation of the activities of the Program for the reconstruction of fixed assets of the complexes, subject to the condition of reducing the total downtime coefficient of all their facilities.

One of the main tools for managing resources of the communications facility is the Program for renewal of fixed assets of communications facilities of JSC Russian Railways.

When preparing proposals on the inclusion of a complex of communication facilities in the Program for the renewal of fixed assets, the ratio of the acceptable, design and factual values of the service availability coefficient is considered as a dependability criterion.

The service availability coefficient characterizes the probability that the requested communication network service will be provided at any time, except for the planned periods during which the use of this service is not expected.

We calculate the actual and design values of the service availability coefficient for each communication facility (in relation to which the issue of renewal is considered), and compare it with the acceptable value of the indicator.

Compliance with the criterion of economic feasibility lies in the fact that when we make a decision on economic feasibility, we perform an economic assessment of the consequences of the decision. The economic assessment consists in comparing the specific average annual costs of ownership of the communication facility (average annual LCC) for various options.

When making decisions on the inclusion of activities in the preliminary list of the Program for the renewal of fixed assets of the complex of communication facilities, we calculate the LCC of the facility for two options: for the case of implementation of the activity and for the case of refusal to implement the activity (alternative option).

We calculate the alternative implementation options (if any) for each project and assess the average annual LCC according to the formula (7.11).

The economic effect of each activity included in the preliminary list of the Program for the renewal of fixed assets of the complex of communication facilities (due to the expired service life and the possibility of obtaining additional income) is determined by the difference between the average annual cost of the facility life cycle when implementing the alternative option and the average annual cost of the facility life cycle when implementing the option. It is calculated by formula (7.12).

The economic effect of the general investment program for the communications complex of JSC Russian Railways (for a given amount of investment) is the achievement of the maximum total economic effect of the implementation of the activities of the Program for the renewal of fixed assets. It is subject to the condition of growth of the total service availability coefficient for all facilities of the communication complex, taking into account the optimal allocation of the activities of the renewal program over time.

## 7.7 Algorithm for Managing the Technical Maintenance of a Facility Based on Risk Assessment Using the Example of a Track Section

The cost of the technical maintenance of the track complex facilities reaches 80% of the cost of the technical maintenance of all infrastructure facilities. For this reason, it is of great practical importance to implement a rational algorithm for managing the technical maintenance of the track complex facilities.

Figure 7.7 shows the algorithm scheme for managing the technical maintenance of a track section; yellow circles show the steps of the algorithm implementation. In some cases these steps are performed in two or three stages.

Let us consider a step-by-step implementation of the algorithm.

*Step 1. Selection of the track maintenance department (TMDept) and open line.* We select the TMDept of the regional directorate and the open line of TMDept in respect of which we will make a decision on the technical maintenance of the track section. This decision is made on the basis of technical and economic indicators. *Step 1.1 Splitting the open line into track sections.* We calculate technical and economic indicators for the track section assessed. For this purpose, the open line is divided into sections according to the mileage. The track section from the kilometer of the beginning of the section to the kilometer of the end of the section is indicated. Recommendations on the technical maintenance are prepared for each of the assessed sections of the open line.

*Step 2. Estimation of the factual tonnage passed.* We estimate the factual tonnage passed through the assessed section according to data from the Unified Corporate Automated System for Infrastructure Management (UC ASUI). The legend of this indicator is $T_{\mathrm{factual}}$.

*Step 3. Comparison of the value of the factual tonnage passed with the regulatory value.* We compare the value of the factual tonnage passed ($T_{\mathrm{factual}}$, mln gross tons (year)) with the regulatory value ($T_{\mathrm{reg}}$, mln gross tons (year)) for the section assessed. If the value $T_{\mathrm{factual}} < T_{\mathrm{reg}}$, then we calculate indicators of operational dependability (*step 3.1*). If the value $T_{\mathrm{factual}} \geq T_{\mathrm{reg}}$, we calculate technical indicators (*step 3.2*). The value $T_{\mathrm{reg}}$ is set by the order of JSC "Russian Railways" No. 75r dated January 18, 2013 (the current version with amendments and additions). The dimension of the indicators is selected depending on the class of the track according to Table 7.5.

*Step 3.1. Calculation of operational dependability indicators.* We calculate key dependability indicators such as the failure rate ($f_i$, failures per year/km) and the availability coefficient ($C_{\mathrm{AV}i}$). The calculation of the values of these dependability indicators is carried out in accordance with the works [11, 12, 13, etc.]. The calculation is automatically made in the Unified Corporate Platform URRAN Track (UCP URRAN T). Next, we proceed to step 9, skipping steps 3.2–8.

*Step 3.2. Calculation of technical indicators characterizing the residual life.* We assess the residual life ($T_{res}$, million gross tons (year)), the predicted value of life before assigning repair ($T_{rep}$) and the expected value of the tonnage passed or service

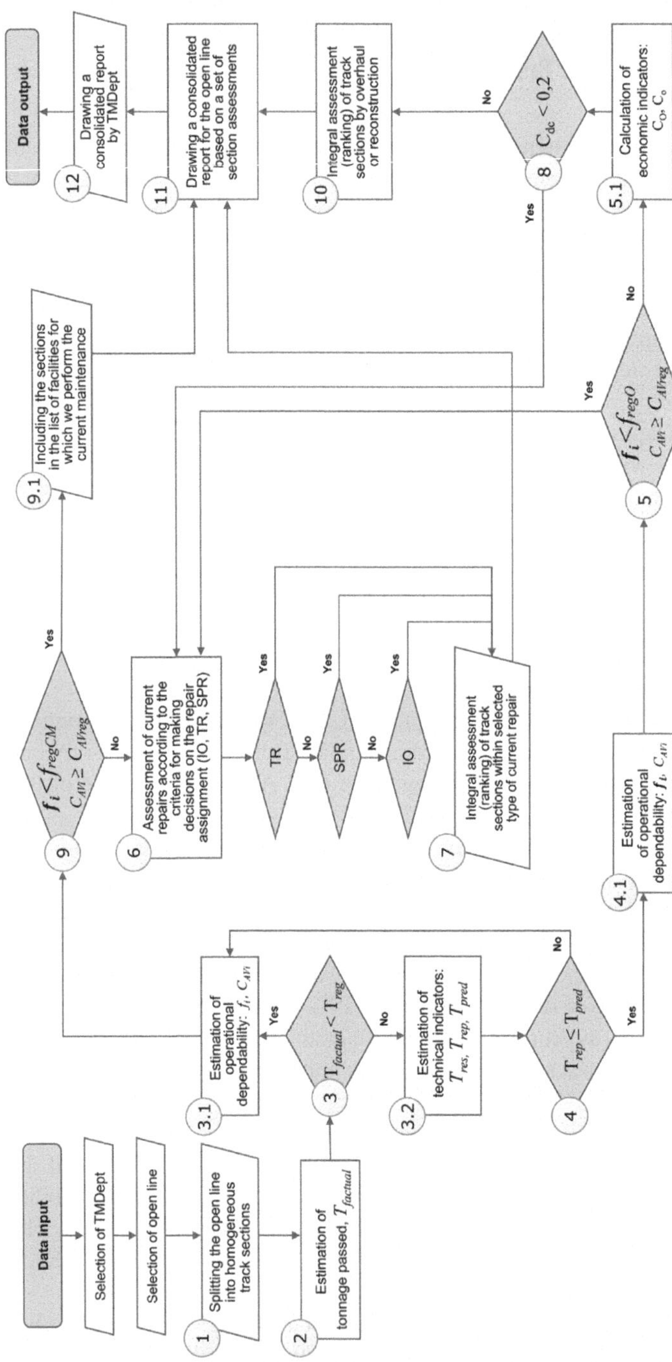


**Fig. 7.7** Algorithm for managing the technical maintenance of a track section

**Table 7.5** Measurement unit of life of a railway track section

| Group code (category) | Track class | Measurement units of life of a railway track section | | | |
|---|---|---|---|---|---|
| | | Concrete sleepers | | Wooden sleepers | |
| | | New materials | Used materials | New materials | Used materials |
| hs (high-speed), s (speed) | 1 | mln gross tons/year | – | – | – |
| p (passenger) | 1, 2, 3 | mln gross tons/year | – | mln gross tons/year | – |
| | 3 | – | Year | – | Year |
| ht (heavy-traffic) | 1, 2 | mln gross tons | – | mln gross tons | – |
| freight I | 1, 2 | mln gross tons | – | mln gross tons | – |
| freight II | 1, 2 | mln gross tons/year | – | mln gross tons/year | – |
| | 3 | – | mln gross tons/year | – | Year |
| freight III, IV, V | 1, 2 | mln gross tons/year | – | mln gross tons/year | – |
| | 3, 3c | – | Year | – | Year |
| | 4, 4c | – | Year | – | Year |
| | 5c | – | Year | – | Year |

life of the facility assessed ($T_{pred}$, million gross tons (year)) for the planned period. The calculation of the values of these indicators is carried out in accordance with STO RZD 08.023-2015 standard (automatically done in the UCP URRAN T). Next, we proceed to step 4.

*Step 4. Comparison of the calculated values of indicators characterizing the residual life*. We compare the values of the predicted life before assigning repair ($T_{rep}$, million gross tons (year)) and the value of expected tonnage passed or service life ($T_{pred}$, million gross tons (year)) for the planned period. If $T_{rep} \leq T_{pred}$, then we proceed to step 4.1. If $T_{rep} > T_{pred}$, then we proceed to step 3.1. The dimension of the indicators is selected depending on the class of the track according to Table 7.5.

*Step 4.1. Calculation of indicators of operational dependability if $T_{rep} \leq T_{pred}$.* Step 4.1 is equivalent to step 3.1. When passing through this algorithm branch, we proceed to step 5.

*Step 5. Comparison of indicators of operational dependability to assign a overhaul (O) or reconstruction*. We compare the values of operational dependability indicators: failure rate ($f_i$, failures per year/km) and availability coefficient ($C_{AVi}$). If $f_i$<$_{\text{freg}}$, and $C_{AVi} \geq C_{AVreg}$, then we proceed to the additional analysis of the facility using the integral assessment (step 6). If $f_i > f_{reg}$ f$_{\text{нормКР}}$ and $C_{AVi} < C_{AVreg}$, then we proceed to the calculation of economic indicators (step 5.1). *Step 5.1. Calculation of economic indicators of the facility assessed.* The coefficient of direct costs $C_{dc}$ is calculated based on the statistics of direct costs of the track current maintenance for the last 2 years and the forecast of overhaul (reconstruction) cost. Next, we proceed to step 8, skipping steps 6–7.1.

*Step 6. Assessment of current repairs according to the criteria for making decisions on the repair assignment.* In accordance with Technical Specifications No. 75r dated January 18, 2013, we assess the criteria for making decisions on the assignment of intermediate overhaul, track raising, and scheduled preventive repair (track alignment). For this purpose, we evaluate each section of the track against specified criteria to identify whether there is a necessity to assign track raising (TR). If, according to these criteria, track raising is not recommended, then we assess track section against the criteria for assigning scheduled preventive repair (SPR). Further, if, according to these criteria, scheduled preventive repair is not recommended, then we assess track section against the criteria for assigning intermediate overhaul (IO).

*Step 7. Integral assessment of track sections within selected type of current repair.* In accordance with the integral assessment, we rank the track sections assessed by priority within each type of repair.

*Step 8. Comparison of the economic indicator values.* The value $C_{dc}$ calculated at step 5.1 is compared with the normative one established in Technical Specifications No. 75rNo. 75r (the normative value is 0.5 as amended on January 19, 2018). If the value exceeds the normative value, then we proceed to step 10, if $C_{dc}$ is less than or equal to the normative value, then we proceed to step 6.

*Step 9. Comparison of operational dependability indicators.* Step 9 is equivalent to step 5. We compare the values of operational dependability indicators, failure rate ($f_i$, failures per year/km), and availability coefficient ($C_{\mathrm{AV}}$). If $f_i$f$_\mathrm{i}$< $f_{\mathrm{reg}}$ $\mathrm{f_{нормТР}}$ and $C_{\mathrm{AV}i} \geq C_{\mathrm{AVreg}}$, then we proceed to the inclusion in the list of sections for which we perform the current maintenance (step 9.1). If $f_i > f_{\mathrm{reg}}$ and $C_{\mathrm{AV}i} < C_{\mathrm{AVreg}}$, then we proceed to the additional analysis of the facility (see step 6). *Step 9.1. Including the sections in the list of facilities for which we perform the current maintenance.* Based on the results of the assessment of operational dependability indicators, we make a decision on satisfactory condition of track section. Next, we proceed to step 11.

*Step 10. Integral assessment (ranking) of track sections that are subject to overhaul and reconstruction.* We rank track sections (open line) by assessing the integral risk (Fig. 7.8).

For this purpose, we assess the risk levels of the following factors:

- Number of defective and flawed rail per 1 track km
- Single withdrawal of defective rails
- Number of defective clamps per 1 track km
- Number of pumping sleepers per 1 track km
- Number of faulty wooden sleepers per 1 track km
- Specific number of places of temporary repair
- Defects of roadbed.

Based on the constructed risk matrices by the method described in item 5.4, an integral risk matrix is formed (Fig. 7.8) for the list of sections. Based on the integral assessment each section gets a priority for its including in itemized list of track overhaul.

In the example shown in the integral matrix of risk of safety violations on track sections 1, 2, 3... $M$, section 2 is of the highest priority (its integral risk is assessed as

| Risk factors | Section 1 | Section 2 | Section 3... | Section M |
|---|---|---|---|---|
| Defects of roadbed | 0,13 | 0,53 | 0,13 | 0,07 |
| Single withdrawal of defective rails | 0,53 | 0,13 | 0,13 | 0,07 |
| Withdrawal of flawed rails | 0,07 | 0,27 | 0,07 | 0,13 |
| Number of defective clamps,% | 0,13 | 0,27 | 0,27 | 0,07 |
| Number of pumping sleepers,% | 0,27 | 0,53 | 0,13 | 0,07 |
| Number of faulty wooden sleepers,% | 0,27 | 0,13 | 0,07 | 0,07 |
| Specific number of places of temporary repair | 0,07 | 0,53 | 0,13 | 0,07 |
| TOTAL | 0,45 | 0,64 | 0,25 | 0,14 |
| Section priority order | 2 | 1 | 3 | 3 |

**Fig. 7.8** Integral matrix of the track section risks

"unacceptable"). This section is ranked with number 1. The next section according to ranking is section 1, then section 3, and so on.

*Step 11. Drawing a consolidated report for each assessed section of the open line.* Data output form at track maintenance department level:

Open line:____________________________

| Section | $T_{res}$ | $f_i$ / $C_{AV}$ | $C_O$ | Recommended type of repair | Order of priority of the section within selected repair type |
|---|---|---|---|---|---|

We record the selected type of repair in the column "Recommended type of repair": SPR, TR, IO, overhaul, reconstruction, or current maintenance.

*Step 12. Drawing a consolidated report by TMDept at the level of the regional directorate.* Data output form (at the regional directorate level):

Regional Directorate

| TMDept | Section | Recommended type of repair | Order of priority of the section within selected repair type |
|---|---|---|---|

## References


1. STO RZD 02.037–2011 Upravlenie resursami na etapah zhiznennogo cikla, riskami i analizom nadezhnosti (URRAN). Upravlenie stoimost'yu zhiznennogo cikla sistem, ustrojstv i oborudovaniya hozyajstv OAO "RZD" (Management of resources at life cycle stages, risks, dependability analysis (URRAN). Management of the life cycle cost of systems, devices and equipment of the facilities of JSC "Russian Railways")
2. Zoeteman, A.: Railway Design and Maintenance from a Life-Cycle Cost Perspective: A Decision-Support Approach (2004)
3. Andrade, A.R.: Renewal decisions from a Life-cycle Cost (LCC) Perspective in Railway Infrastructure: An integrative approach using separate LCC models for rail and ballast components (2008)
4. GOST 33433-2015 Mezhgosudarstvennyj standart "Bezopasnost' funkcional'naya. Upravlenie riskami na zheleznodorozhnom transporte" (Interstate standard "Functional safety. Risk management in railway transportation")
5. Sedyakin, N.M.: Ob odnom fizicheskom principe teorii nadezhnosti i nekotoryh ego prilozheniyah (On a physical principle of dependability theory and some applications of this principle). LVIKA im. A.F. Mozhajskogo, Leningrad (1965)
6. Sedyakin, N.M.: Ob odnom fizicheskom principe teorii nadezhnosti (On a physical principle of the dependability theory). Izv. AN SSSR. Tekhnicheskaya kibernetika. **3**, 80–87 (1966)
7. GOST 32192-2013 Mezhgosudarstvennyj standart "Nadezhnost' v zheleznodorozhnoj tekhnike. Osnovnye ponyatiya. Terminy i opredeleniya" (Interstate standard "Dependability in railway technics. General concepts. Terms and definitions")
8. Metodika ocenki effektivnosti prodleniya sroka sluzhby osnovnyh sredstv hozyajstva svyazi OAO "RZD" na osnove metodologii URRAN (Method for assessing the effectiveness of prolongation of the service life of communication complex fixed assets of JSC "Russian Railways" based on the URRAN methodology), Moscow, RZD (2014)
9. Metodika ocenki funkcional'nogo resursa tekhnicheskih sredstv zheleznodorozhnoj avtomatiki i telemekhaniki (Method for assessing the functional life of technical means of signalling and remote control facilities), Moscow, RZD (2015)
10. Belyaev, Y.K., et al.: Nadezhnost' tekhnicheskih sistem: Spravochnik. In: Ushakov, I.A. (ed.) Reliability of Technical Systems: Handbook. Radio i svyaz', Moscow (1985)
11. IEC 62278:2002 Railway applications – Specification and demonstration of reliability, availability, maintainability and safety (RAMS)
12. Metodika rascheta pokazatelej nadezhnosti i bezopasnosti funkcionirovaniya etalonnyh ob"ektov putevogo hozyajstva OAO "RZD" (Method for calculating dependability and safety indicators of operation of the reference objects of JSC "Russian Railways" track facilities), Moscow, RZD (2011)
13. Kumar, S.: Reliability Analysis and Cost Modeling of Degrading Systems. Division of Operation and Maintenance Engineering, Luleå University of Technology (2008)

# Chapter 8
# Assessment of the Activities of Structural Divisions of Railway Transport

## 8.1 General Provisions

Assessment of the activities of structural divisions of railway transport is one of the most important technologies for managing the transportation process. The results of the assessment make it possible to define the effectiveness of the division activities regarding fulfillment of production tasks, ensuring the safety and dependability of train traffic, ensuring prompt restoration of the division's facility operability, and activities to ensure the required level of competence of the division's personnel. The results of the assessment are a key tool for managing the activities of divisions, branches, and directorates of the Russian Railways as well as for assigning investments to subdivisions.

Ranking and comparative assessment methods provide the most convenient and practical form of results presentation. The presentation of results in the form of scores in combination with a qualitative rating scale is one of the forms that has proven itself in many areas of activity.

Regarding structural divisions of infrastructure complexes it is customary to assign scores using a 100-point scale. The scale provides for four ranking categories: "excellent," "good," "satisfactory," and "unsatisfactory" (Fig. 8.1).

Regarding structural divisions of the *locomotive complex* (locomotive depots), it is customary to assign scores using 7000-point negative scale. The scale provides for four assessment categories: green zone—"dependability zone," yellow zone—"development zone," orange zone—"significant risk zone," and red zone—"critical risk zone" (Fig. 8.2).

It is possible to assess activities of the division by expertise (for example, based on the results of an audit) or even by the number of facility equipment failures in the division. However, such approaches are far from objective assessment. A simple example. Let two track maintenance departments work in the same climatic conditions, and there are 2 times more failures in track maintenance department No. 1 than

 
I. B. Shubinsky, A. M. Zamyshlaev, *Technical Asset Management for Railway Transport*, International Series in Operations Research & Management Science 322,
https://doi.org/10.1007/978-3-030-90029-8_8

| The value of the indicator, scores (structural divisions of the track complex and the CCS) | The value of the indicator, scores (structural divisions of signalling and remote control facility) | Score assessment categories |
|---|---|---|
| 0–25 | ≤ 80 | "Excellent " |
| 26–50 | ≤ 60 - <80 | "Good" |
| 51–75 | ≤ 40 - <60 | "Satisfactorily" |
| 76–100 | < 40 | "Unsatisfactory" |

**Fig. 8.1** Score assessment categories for infrastructure divisions

| Indicator value, scores | Rating category |
|---|---|
| >1200 | "Dependability zone" - excellent |
| ≥ 1201 - ≤ 2499 | "Development zone" - good |
| >2500 - ≤ 6499 | "Significant risk zone" - satisfactory |
| >7000 | "Critical risk zone" - unsatisfactory |

**Fig. 8.2** Score rating categories for the locomotive depot

in track maintenance department No. 2 for a given observation period. At first glance, the ranking position of the track maintenance department No. 2 is higher than the ranking position of the track maintenance department No. 1. However, it appears that the volume of freight traffic handled by the first track maintenance department is 1.5 times more than the volume of freight traffic handled by the second track maintenance department for a given observation period. If we convert the number of failures into dependability indicators taking into account the load intensity, the results of the assessment will be opposite—the track maintenance department No. 1 will take a higher ranking position than the track maintenance department No. 2. This example is very simplistic. In fact, one should take into account the quality of activities performed by divisions (promptness of malfunction repair, ensuring the safety and dependability of the transportation process on the basis of a risk assessment). The qualification and training of personnel should be also taken into account. The conditions (climatic conditions, equipment capabilities, amount of traffic, etc.) under which they perform their tasks have a great influence on the comparative assessment of division activities.

The URRAN methodology allows combining expert assessment with statistical data on safety, operational dependability, and promptness of recovery. At the same time, it allows taking into account the risks of violation of safety and dependability, and, of course, climatic and other conditions under which divisions perform production tasks. The latter circumstance is taken into account by standardizing facilities (Chap. 5).

## 8.2 Assessment of the Activities of Track Complex Divisions

The purpose of assessing the activities of the structural divisions of the track complex is to obtain an objective assessment based on the analysis of their activities on the railway network.

As the subject of assessment we take the results of the activities of a separate structural division of the track complex for a certain observation period. It is recommended to choose the observation interval equal to 1 year (8760 h). At the same time, we should ensure a consistent approach to assessing the activities of track maintenance department or the track maintenance division of the regional infrastructure directorate. The assessment is performed with a given level of detail.

We assess the activities of the structural divisions of the track complex of the Russian Railways on the basis of a multilayer set of criteria that characterize the key aspects of the track complex activities. Integral assessment allows determining structural divisions' activities on the basis of a set of key indicators that help to assess the activities of this division in certain areas. Key indicators, in turn, are determined on the basis of local (primary) indicators. We use the following key indicators to obtain an integral assessment SI of the activities of structural divisions of the track complex:

- Scoring train traffic safety ($S^{\mathrm{TS}}$)
- Scoring dependability ($S^{\mathrm{D}}$)
- Scoring personnel competence ($S^{\mathrm{PC}}$).

The assessment of the activities of the structural division of the track facilities for a given observation period should be carried out in three stages:

- At the first stage, we perform score assessment of train traffic safety ($S^{\mathrm{TS}}$). If the result of the assessment is unsatisfactory, then the further stages of the assessment are not carried out. In this case the final integral assessment of the division is considered as "unsatisfactory"
- At the second stage, we perform score assessment of the dependability of the track complex facility ($S^{\mathrm{D}}$). It comprises three components: score assessment of the quality of the technical maintenance ($S^{\lambda}$), score assessment of promptness of upper structure failure elimination ($S^{\mathrm{R}}$), and score assessment of the impact on the transportation process ($S^{\mathrm{TP}}$)
- At the third stage, we assess the competence of the personnel ($S^{\mathrm{PC}}$).

Further, the results of the assessments are combined to obtain an integral assessment. Also the ranking of the divisions is compiled.

The score assessment of ensuring the safety of train movements is calculated on the basis of the factual level of risk $R$ (for all types of train traffic safety violations [1]), the corresponding acceptable level of risk $R_{\mathrm{acceptable}}$ and the coefficient $C$ of the risk matrix.

We can use the following criterion when there is no information on risk.

Taking into account the fact that traffic safety violations are rare (especially if the subject of the assessment is not the entire network, but the network of the regional directorate of infrastructure or track maintenance department), it is possible to normalize the acceptable (regulatory) number $m_{reg}$ of such events using the upper confidence limit of the Poisson distribution:

$$m_{reg} = \{r_0, m_{av} = 0; | \tag{8.1}$$

where:

$m_{av}$ is the average value of the number of traffic safety violations for which the assessed structural division is responsible; this value is rounded off to the nearest whole number

$r_0$, $r(m_{av})$ is the tabular values of the upper confidence limits of the Poisson distribution for a given confidence probability (it is recommended to set a confidence probability equal to 0.9).

The score assessment of ***train traffic safety*** provided by structural division is carried out according to the following formula:

$$S^{TS} = \max\left\{0; \min\left\{100; 75 - \frac{25\lg\left(\frac{R}{R_{accepable}}\right)}{\lg(C)}\right\}\right\} \tag{8.2}$$

where:

$R$ is the factual level of risk associated with traffic safety violations (of all types)

$R_{accepable}$ is the acceptable level of risk associated with traffic safety violations (of all types)

$C$ is the scale coefficient of the risk matrix associated with traffic safety violations.

If there is no information on risk, the scoring is performed using the following formula:

$$S_k^{TS} = \min\left\{100; 75\left(\frac{m}{m_{reg}}\right)^{\eta}\right\} \tag{8.3}$$

where:

$m_{reg}$ is the regulatory value of the number of traffic safety violations for which structural division is responsible (formula (8.1))

$m$ is the number of traffic safety violations for which structural division is responsible (for the considered observation period)

η is a correction indicator of the traffic safety scale (it is recommended that $\eta =$ 1.0 for practical calculations).

A score assessment of the ***dependability of infrastructure facilities*** of the track complex begins with an *assessment of the quality of the technical maintenance* of infrastructure facilities within structural division. It is done using the indicator $S^{\lambda}$, which is calculated on the basis of the ratio of the factual rate of the track upper structure failure (within the responsibility of a given structural division) and the regulatory value of the track upper structure failure rate. In this case, we apply the values of the mentioned indicators, reduced to a section of track with a length of 10 km.

The factual specific failure rate $\lambda_K$ is calculated for each *k*th group of homogeneous facilities of the upper structure sections within the responsibility of a given structural division. The regulatory values of the specific failure rate $\lambda_{reg}$, mean time to recovery $T_{rec\ reg}$, and downtime coefficient $\widehat{C}_{d\ reg}$ are set for all groups of homogeneous facilities in accordance with [1].

On the basis of the values $\widehat{\lambda}_K$ and $\lambda_{reg}$ we perform the score assessment of the quality of the facility technical maintenance of the selected *k*th group of homogeneous facilities:

$$S_k^{\lambda} = \min\left\{100; 75\left(\frac{\widehat{\lambda}_K}{\widehat{\lambda}_{reg}}\right)^{\alpha}\right\} \tag{8.4}$$

where α is the correction indicator of the technical maintenance quality scale (it is recommended that $\alpha = 3$ for practical application).

The ratio of the factual specific failure rate to the regulatory one makes it possible to assess the extent to which the quality of the technical maintenance of infrastructure facilities allows ensuring the level of technical means dependability set in the technical specification and the design projects.

We evaluate *the promptness of elimination of infrastructure facilities failures* within structural division using an indicator $S^R$. It is calculated on the basis of the ratio of the actual average time to recovery of the upper structure of the track (within the responsibility of the given structural division) to the regulatory value of the average time to recovery of upper structure of the track.

The actual average time to recovery is $T_{Rk}$calculated for each *k*th group of homogeneous objects of the upper structure of the path on the sections within the responsibility of the specified structural division.

We perform score assessment of the promptness of eliminating facility failures for the selected *k*th group of homogeneous facilities based on the values $T_{Rk}$ and $T_{R\ reg}$:

$$S_k^R = \min\left\{100; 75\left(\frac{T_{Rk}}{T_{R\ reg}}\right)^{\beta}\right\} \tag{8.5}$$

where $\beta$ is the correction indicator of the scale of the failure elimination promptness (it is recommended that $\beta = 3$ for practical application).

The ratio of the factual average time to recovery to the normative one makes it possible to assess how the promptness of the elimination of infrastructure facilities failures allows ensuring the level of technical means dependability set in the technical conditions and in the developed projects.

We assess *the influence of infrastructure facilities on the transportation process* within structural division using an indicator $S^{\mathrm{TP}}$. This indicator is calculated on the basis of the ratio of the factual coefficient of downtime of the upper structure of the path (within the scope of the responsibility of the given structural division) to the regulatory value of the coefficient of downtime of the upper structure of the path. In this case, we apply the values of the mentioned indicators normalized to a track section with a length of 10 km. The factual specific downtime coefficient $\widehat{C}_{dk}$ is calculated for each *k*th group of homogeneous facilities of the track upper structure within the sections for which structural division is responsible.

Based on the values $\widehat{C}_{dk}$ and $\widehat{C}_{d\ \mathrm{reg}}$, we perform the score assessment of the influence of facilities (for selected *k*th group of homogeneous facilities) on the transportation process:

$$S_k^{\mathrm{TP}} = \min\left\{100; 75\left(\frac{\widehat{C}_{dk}}{\widehat{C}_{d\ \mathrm{reg}}}\right)^{\gamma}\right\} \tag{8.6}$$

where $\gamma$ is a correction indicator of the scale of the influence on the transportation process (it is recommended that $\gamma = 3$ for practical application).

The ratio of the factual specific downtime coefficient to the regulatory one makes it possible to assess to what extent the division ensures the fulfillment of requirements for the uninterrupted transportation process.

Based on the scores obtained for each group of homogeneous facilities, we calculate the final scores for the technical maintenance quality $S^{\lambda}$, the promptness of eliminating failures $S^{\mathrm{R}}$, and the influence on the transportation process $S^{\mathrm{TP}}$.

$$S^{\lambda} = \frac{1}{N}\sum_{k=1}^{N} S_k^{\lambda} \quad S^{R} = \frac{1}{N}\sum_{k=1}^{N} S_k^{R} \quad S^{\mathrm{TP}} = \frac{1}{N}\sum_{k=1}^{N} S_k^{\mathrm{TP}} \tag{8.7}$$

where N is the number of groups of homogeneous facilities.

The final score assessment of the level of dependability of infrastructure facilities within division is formed on the basis of score assessments of the quality of technical maintenance ($S^{\lambda}$), the promptness of eliminating failures ($S^{r}$), and the influence on the transportation process ($S^{\mathrm{tp}}$). At the same time, the priority indicator is the score assessment of the influence on the transportation process. The calculation is performed according to the following formula:

$$S^{D} = 0.3S^{\lambda} + 0.3S^{R} + 0.4S^{\mathrm{TP}} \tag{8.8}$$

We assess ***the competence of the personnel of the structural division*** using the indicator $S^{\mathrm{PC}}$, which is calculated on the basis of comparing the factual share of facility failures number (due to personnel errors) with its regulatory value $C_{\mathrm{pc\ reg}}$.

The score assessment of the competence of the structural division personnel is performed using data on the factual number *n* of facility failures and the factual number $n_{pc}$ of failures that occurred due to personnel errors ($n_{pc} \leq n$) over the observation interval:

- in case when $n = 0$, we assume $S^{PC} = 0$;
- in case when $n_{pc} < 8$ and $n \geq \frac{1.68}{C_{pc\ reg}}$, as well as when $n_{pc} \geq 8$ and $n > 0$:

$$S^{PC} = \min\left\{100; \frac{50\ n_{pc}}{n\ C_{pc\ reg}}\right\}, \tag{8.9}$$

where:

$n_{pc}$ is the number of facility failures (by division) due to personnel errors for the observation interval

$n$ is the total number of facility failures (by division) for the observation interval

$C_{pc\ reg}$ is the regulatory value of the share of failure number due to personnel errors (based on the research findings, it is recommended that $C_{pc\ reg} = 0.17$)

- in case when $0 < n < \frac{1.68}{C_{pc\ reg}}$:

$$S^{PC} = \min\left\{100; 50A\left(n\ C_{pc\ reg} - B\right)^2 + C\right\} \tag{8.10}$$

where *A*, *B*, and *C* are coefficients depending on $n_{pc}$ and given in [2].

As for formulas (8.2)–(8.6) and (8.9)–(8.10) it is provided that the level of scores of a key indicator cannot exceed 100 scores. A comprehensive (integral) assessment of the activity of a structural division is formed on the basis of the scores $S^{TS}$, $S^{D}$, and $S^{PC}$. In this case, the priority is given to the indicator of ensuring traffic safety. The level of scores of the comprehensive assessment cannot be lower than the number of scores of the $S^{TS}$. Therefore, the calculation of the integral indicator of activities is carried out according to the following formula:

$$S^{D} = \max\left\{S^{TS}; (0.4S^{TS} + 0.3S^{D} + 0.3S^{PC}\right\}. \tag{8.11}$$

Table 8.1 on the basis of integral score assessments ($S^{I}$), it is possible to make a comparative assessment of the activities of various structural divisions of track complex, as well as the dynamics of the quality of the activities of the considered

**Table 8.1** Coefficients *A*, *B*, and *C* for formula (8.10)

| $n_{pc}$ | *A* | *B* | *C* |
|---|---|---|---|
| 0 | 0.17715 | 1.68 | 0 |
| 1 | -0.03374 | -0.94511 | 0.53014 |
| 2 | -0.24463 | 0.95582 | 0.72349 |
| 3 | -0.45554 | 1.09667 | 1.04787 |
| 4 | -0.66644 | 1.14835 | 1.37884 |
| 5 | -0.87734 | 1.17519 | 1.71167 |
| 6 | -1.08823 | 1.19163 | 2.04527 |
| 7 | -1.29913 | 1.20273 | 2.37926 |

divisions during several adjacent assessment periods. To do this, we should use the activities ranking method.

The principle of the ranking method lies in the fact that the structural division occupies a certain ranking position among divisions depending on the value of the comprehensive (integral) score assessment for the selected assessment period. The less score given division obtains, the higher position in the list it occupies (it is ranked in ascending score order). Ranking positions are assigned from the first to the last position. The ranking can be done not only by comprehensive indicators, but also by key indicators $S^{\mathrm{TS}}$, $S^{\mathrm{D}}$, or $S^{\mathrm{TP}}$. Thus, the higher ranking position structural division occupies, the higher quality of its activities is for the assessment period (in comparison with other structural divisions of the track complex); and the lower ranking position it occupies, the worse the quality of its activities is. Comparing the ranking position of the division for adjacent periods of assessment, it is possible to assess the dynamics of the quality of its activities. For example, we can assess whether there is a decrease or increase in the quality of division activities in comparison with the quality of other division activities.

## 8.3 Assessment of the Activities of Structural Division of Central Communications Station

*The Central Communications Station (CCS)* includes the following structural divisions: communications directorate and regional communications centers. Their main purpose is to organize uninterrupted operation of communication networks and to maintain the structures, devices, and communication equipment in good state within the scope of their responsibility.

A 100-point scale is used as an assessment scale for ranking the activities of railway telecommunication divisions (see Fig. 8.1).

Table 8.2 gives the list of primary indicators, activities assessment criteria, and final results according to which we perform a score assessment of the activities of *CCS* structural divisions.

***A set of particular indicators*** (for assessing the activities of CCS structural divisions) is formed on the basis of primary indicators of dependability and quality criteria (see Table 8.2) and includes the following eight indicators:

- Indicator $I_{11}$ of unavailability of railway telecommunication facilities
- Indicator $I_{121}$ of the failure rate of railway telecommunication facilities
- Indicator $I_{122}$ of the share of pre-failures of railway telecommunication facilities
- indicator $I_{13}$ of the average time to recovery of railway telecommunication facilities
- Indicator $I_{21}$ of unavailability of railway telecommunication services, the failures of which led to train delays
- Indicator $I_{22}$ of the failure rate of railway telecommunication services that leads to train delays

**Table 8.2** Indicators, assessment criteria, and activity results of the communication structural division

| Activity results | Assessment criteria | Primary indicators of dependability | Factors taken into account |
|---|---|---|---|
| 1. Sustainability of the functioning of railway telecommunications | 1.1 Quality of functioning of facilities | 1.1 The coefficient of unavailability of railway telecommunication facilities | a) State of technical means.<br>b) Climatic operating conditions.<br>c) Availability of TM and R tools. |
| | 1.2. Quality of the technical maintenance of facilities | 1.2.1 Failure rates of railway telecommunication facilities<br>1.2.2 The share of pre-failures of railway telecommunication facilities | |
| | 1.3. Promptness of elimination of facility failures | 1.3.1 Average time to recovery of railway telecommunication facilities | |
| 2. Influence of the functioning of railway telecommunications on the transportation process | 2.1. The quality of services, the failures of which led to train delays | 2.1.1. The coefficient of unavailability of services, the failures of which led to train delays | a) State of technical means.<br>b) Climatic operating conditions.<br>c) Availability of TM and R tools. |
| | 2.2 Influence of technical maintenance on service failures that led to train delays | 2.2.1. Failure rate of services provided that the failure of this services led to train delays | |
| | 2.3. Promptness of elimination of service failures that led to train delays | 2.3.1. Average time to recovery of services, failure of which led to train delays | |

- Indicator $I_{23}$ of average time to recovery of railway telecommunication services, the failures of which led to train delays
- Indicator $I_{31}$ of the level of the operating personnel competence.

The listed indicators are measured using scores in accordance with the general formula:

$$I_r = \min\left\{100; 75(f_i)^{\beta_i}\right\} \tag{8.12}$$

where $\beta_i$ is the correction index for calculating the $i$th indicator.

In the formula (8.12) *in respect to the indicator* $I_{11}$, the function

$$(f_{11})^{\beta_{11}} = \left(\frac{i_{\mathrm{UF}}}{i_{\mathrm{UF}}^{\mathrm{acceptable}}\, i_{\mathrm{Resp}}}\right)^{0.7}$$

where $i_{\mathrm{UF}}$ is the actual coefficient of facility unavailability (within division); $i_{\mathrm{UF}}^{\mathrm{acceptable}}$ is the acceptable coefficient of facility unavailability (within the whole network); $i_{\mathrm{Resp}}$ is a coefficient that takes into account the division responsibility for ensuring failure-free operation of railway telecommunications.

In formula (8.12) *in respect to the indicator* $I_{121}$, the function

$$(f_{121})^{\beta_{121}} = \left(\frac{\lambda}{\lambda^{\mathrm{acceptable}}\, i_{\mathrm{Resp}}}\right)^{1}$$

where:

$l$ is the actual rate of facility failures (within division) for the observation interval, 1/h

$\lambda^{\mathrm{acceptable}}$ is the acceptable rate of facility failure (within the whole network), 1/h.

In formula (8.12) *in respect to the indicator* $I_{122}$, the function

$$(f_{122})^{\beta_{122}} = \left(\frac{n_{\mathrm{PF}}}{N_{PF}\, i_{\mathrm{Resp}}}\right)^{1.2}$$

where $n_{\mathrm{PF}}$ is the actual number of pre-failures of facilities (that are under control of the division) for the observation interval; $N_{\mathrm{PF}}$ is the actual number of pre-failures throughout the whole railway telecommunication network during the observation interval.

If $N_{\mathrm{PF}} = 0$, then $I_{\mathbf{122}} = 0$. If there are no initial data on pre-failures, then we should assume $I_{122} = 0$.

In formula (8.12) *in respect to the indicator* $I_{13}$, the function

$$(f_{13})^{\beta_{13}} = \left(\frac{T_R}{T_R^{\mathrm{acceptable}}}\right)^{1}$$

where:

$T_R$ is the actual average time to recovery of facilities (within division)

$T_R^{\mathrm{acceptable}}$ is the acceptable average time to recovery of facilities (within the whole network).

In formula (8.12) *in respect to the indicator* $I_{21}$, the function

$$( f_{21})^{\beta_{21}} = \left( \frac{i_{US}}{i_{\mathrm{US}}^{\mathrm{acceptable}} \cdot i_{\mathrm{Resp}}} \right)^{0.7}$$

where $I_{\mathrm{US}}$ is the actual coefficient of unavailability of services, the failures of which led to train delays; $i_{\mathrm{US}}^{\mathrm{acceptable}}$ is the acceptable coefficient of unavailability of services, the failures of which led to train delays (within the whole network).

In formula (8.12) *in respect to the indicator* $K_{22}$, the function

$$( f_{22})^{\beta_{22}} = \left( \frac{\lambda_S}{\lambda_S^{\mathrm{acceptable}} \cdot i_{\mathrm{Resp}}} \right)^{1}$$

where:

$l_S$ is the actual rate of service failures that led to train delays

$\lambda_S^{\mathrm{acceptable}}$ is the acceptable rate of service failures that led to train delays.

In formula (8.12) *in respect to the indicator* $K_{23}$, the function

$$( f_{23})^{\beta_{23}} = \left( \frac{T_{\mathrm{RS}}}{T_{\mathrm{RS}}^{\mathrm{acceptable}}} \right)^{1}$$

where:

$T_{\mathrm{RS}}$ is the actual average time to recovery of services, the failures of which led to train delays (within division) for the observation interval

$T_{\mathrm{RS}}^{\mathrm{acceptable}}$ is the acceptable average time to recovery of services, the failures of which led to train delays.

*The indicator* $I_{31}$ *of the level of competence of the operating personnel* within the assessed division is calculated using the following formulas:

- If $n_P < m$:

$$I_{31} = \begin{cases} \min\left\{100; 50\left(\frac{n_P}{n\ I_P^{\mathrm{acceptable}}}\right)\right\}, n\ I_P^{\mathrm{acceptable}} \geq q; \\ \min\left\{100; A\left(n\ I_P^{\mathrm{acceptable}} - B\right)^2 + C\right\}, 0 < n\ I_P^{\mathrm{acceptable}} < q; \\ 0, n = 0 \end{cases} \quad (8.13)$$

- If $n_P \geq m$ :

$$I_{31} = \begin{cases} \min\left\{100; 50\left(\dfrac{n_P}{n\ I_P^{\text{acceptable}}}\right)\right\}, n > 0; \\ 0, n = 0 \end{cases} \tag{8.14}$$

where:

$n_P$ is the number of facility failures (within division) caused by personnel errors over the observation period; $n$ is the total number of facility failures (within division) over the observation interval

$0.01 \le I_P^{\text{acceptable}}$ is the acceptable share of failures caused by personnel errors (it is recommended that $I_P^{\text{acceptable}} = 0.12$)

$1 \le q < 2$ is the value of the product $n\ I_P^{\text{acceptable}}$, below which the reliability of assessment (of personnel competence level) decreases due to insufficient sample size of statistical data

$m$ is the limiting value, which depends on the lower limit (the recommended value is 0.01) and the specified value $q$, at which the values of function $A\left(n\ I_P^{\text{acceptable}} - B\right)^2 + C$ (for any $m \le n_p < q/I_P^{\text{acceptable}}$) will be greater than 0.75

$A$, $B$, and $C$ are coefficients depending on the values $q$ and $n_p$, which are presented in Table 8.1 for $q = 1.68$ (recommended value; with $m = 8$).

***An integral assessment*** of the activities of CCS structural divisions is performed on the basis of the following four integral indicators:

- Indicator $B_0$ of operating conditions
- Indicator $B_1$ of the sustainability of the functioning of railway telecommunications
- Indicator $B_2$ of the influence of the functioning of railway telecommunications on the transportation process
- Indicator $B_3$ of personnel competence.

*The general integral indicator* $B_\Sigma$ for assessing the division activities is determined on the basis of the integral indicators $B_1 - B_3$.

We calculate these integral indicators *by the results of calculations of particular indicators taking into account the factors of operating conditions* $F_i$ *(i = 1,2,3)*, where $F_1$ *is the factor of the technical means state.* This factor characterizes the level of their wear and tear. As already noted in Sect. 5.4, the overwhelming majority of the equipment of railway telecommunication facilities wears out much more slowly than it becomes obsolete. Therefore, if there are no results of studies on obsolescence and wear of electromechanical equipment, such as cabinets, cables, etc., then this factor can be estimated using the formula recommended in [2]:

$$F_1 = F_{1\max} = 1 + 0.5\alpha_1,$$

where $\alpha_1 = 0.923$ is the normalization coefficient.

*Factor $F_2$ of climatic operation conditions* is determined on the basis of data on the geographical location of the railway line sections maintained by the structural division. It is calculated according to the formula

$$F_2 = \prod_{i=1}^{4} c_{cli}^{v_i}$$

where:

$c_{\mathrm{cl}\ 1} \ldots c_{\mathrm{cl}\ 4}$ are the coefficients of classes 1 ... 4 of climatic operation conditions

$v_i = \frac{l_i}{\sum_{i=1}^{4} l_i}, i = 1 \ldots 4$ is the share of railway line sections maintained by the structural division and located in the climatic conditions zone of the *i*th class, km

$l_i$ is the total length of the railway line sections maintained by the structural division located in the climatic conditions zone of the *i*th class, km.

*The factor $F_3$ of availability of tools for technical maintenance and repair (TM* and *R)* is established on the basis of an expert assessment of the compliance of availability of tools with operation requirements. If the availability of tools meets the operation requirements, it is recommended that $F_3 = 1$, if it does not meet, then it is recommended that $F_3 = 1.3$. If there is no information on the availability of TM and R tools, then we assume that $F_3 = 1$.

*The integral indicator $B_0$ of operating conditions* is calculated by the formula:

$$B_0 = \frac{100(F_1F_2F_3 - 1)}{F_{1\,max}F_{2\,\max}F_{3\,\max} - 1} \tag{8.15}$$

where:

$F_{1\max}$ is the upper limit of the factor $F_1$ ($F_{1\max} = 1 + a_1 = 1.923$)

$F_{2\max}$ is the upper limit of the factor $F_2$ ($F_{2\max} = 1.6$)

$F_{3\max}$ is the upper limit of the factor $F_3$ ($F_{3\max} = 1.3$).

*The integral indicator $B_1$* of the sustainability of railway telecommunications functioning is calculated using the following formula:

$$B_1 = \left(\frac{1}{F_1F_2F_3}\right)^{\gamma_F} \max\{I_{11}I_{121}I_{122}I_{13}\} \tag{8.16}$$

*The integral indicator $B_2$ of the influence of the railway telecommunications functioning on the transportation process within the structural division* is calculated using the following formula:

$$B_2 = \left(\frac{1}{F_1F_2F_3}\right)^{\gamma_F} \max\{I_{21}I_{22}I_{23}\} \tag{8.17}$$

**Table 8.3** Assessment of the integral indicators significance that are used for assessing structural division activities

| Integral indicator | Significance assessment, $\xi$ | Weight of integral indicator |
|---|---|---|
| $B_{2-}$ (the influence of functioning of railway telecommunications on the transportation process) | $\xi_2 = 10$ | $\alpha_2 = \frac{\xi_2}{\xi_1+\xi_2+\xi_3} = 0.435$ |
| $B_1$ (the sustainability of the functioning of railway telecommunications) | $\xi_1 = 7$ | $\alpha_1 = 0.304$ |
| $B_3$ (personnel competence) | $\xi_3 = 6$ | $\alpha_1 = 0.261$ |

*The integral indicator $B_3$ of personnel competence* is calculated using the following formula:

$$B_3 = \left(\frac{1}{F_1 F_2 F_3}\right)^{\gamma_F} I_{31} \tag{8.18}$$

In formulas (8.16), (8.17), and (8.18) the correction index

$$\gamma_F = \frac{\lg(\nu)}{\lg(F_{1\max} F_{2\max} F_{3\max})}.$$

($\nu = 1, 2 \ldots 4$ is an indicator of the influence of operating conditions; it is recommended that $\nu = 2$, hence, if $F_{1\max} = 1.923$, $F_{2\max} = 1.6$, $F_{3\max} = 1.3$, then $\gamma_F = 0.5$).

**The general integral indicator $B_\Sigma$** of the division activity is calculated as the sum of the integral indicators $B_1 - B_3$, taking into account the corresponding weight coefficients $\alpha_1 - \alpha_3$. $B_\Sigma$ is a generalized score assessment of the division activities:

$$B_\Sigma = \sum_{j=1}^{3} \alpha_j B_j \tag{8.19}$$

We *choose weight coefficients for integral indicators of assessment of the structural division activities* on the basis of expert assessments of integral indicators significance (see Table 8.3). It is done to form a general integral indicator.

## 8.4 Assessment of the Activity of Railway Signalling and Remote Control Divisions

### *8.4.1 Introduction*

The basic structural unit of the railway signalling and remote control (SRC) complex is the signalling and interlocking (SI) department. The activity of SI department is assessed on the basis of the results of solving the following main tasks:

- Ensuring the safety of train traffic
- Keeping the SRC equipment in technically good condition within the established scope of responsibility of a department, prevention and elimination of malfunctions of their normal operation, increasing the dependability, efficiency, and cost effectiveness of the SRC equipment of SI department.

The activity of the SI department is assessed on the basis of an integral indicator determined by the results of estimation of the following indicators: the basic indicator and additional indicators of SI department activity.

The basic indicator is determined using the values of the integral quality indicators of technical operation of SRC individual object of SI department.

The integral quality indicator of technical operation of SRC object and SI department as a whole is determined by a comprehensive assessment of the basic and additional quality indicators of the technical operation of the SRC facility. The basic quality indicator of the technical operation of the SRC object and SI department as a whole is determined on the basis of comparing the actual values of the indicators of the operation safety and dependability of the SRC facility with the established standards, and by the influence of the facility dependability on the duration of train delays. There are the following values of the qualitative assessment of the basic indicator of the SI department activity: "excellent," "good," "satisfactory," and "unsatisfactory." Each of these values corresponds to the ranges of quantitative assessments expressed in scores.

The list of additional indicators of the SI department activity as well as the norms for their assessment is set by the order of the Central Infrastructure Directorate of the Russian Railways. They can annually be revised.

If the failure or disruption in the operation of the SRC equipment (due to a fault of the personnel of the SI department) led to an event classified as a transport accident (train crash, accident, and incident at a railway level crossing) or to other events related to violation of requirements for safety of traffic and railway transport operation (except for a train delay of 1 h or more), then we determine the basic assessment of the indicator of SI department activity for a month as "unsatisfactory."

SI department ranking is performed on the basis of a quantitative assessment of the integral indicators of department activities.

### 8.4.2 Assessment of the Quality of SRC Facility Technical Operation by SI Department

We assess the basic indicator of SI department activity using results of assessment of the integral quality indicators of the technical operation of the constituent SRC objects (see (8.20)).

As for the integral indicator of each SRC objects it is determined by summing up the basic and additional indicators of the quality of technical operation. The basic indicator is formed on the basis of the results of assessment of the actual facility dependability level and the predicted value of the risk level due to its undependability. Paper [1] proposes the following table for a quantitative and qualitative assessment of the basic quality indicator of SRC facility technical operation (Fig. 8.3). The quantitative assessment is performed using scores. The qualitative assessment is performed using four ranking categories "Excellent," "Good," "Satisfactory," "Unsatisfactory."

The actual level of dependability is calculated using the statistical data on failures of categories 1 and 2 with the help of the value of the availability coefficient ($K_{av}$). Category 1 failure is a failure that led to a train delay of 60 min or more. In case of category 2 failure, train delay is in the interval of 10–60 min. Taking into account the class and specialization of the railway line, we form 4 ranges of values of the SRC facility availability coefficient. The four levels of the SRC facility dependability correspond to the four levels of the consequences of the SRC facility failure (Fig. 8.4). The levels of the SRC facility dependability are determined by the following inequalities:

1 $K_{av(1-2)} \leq K_{av}$ 2 $K_{av(2-3)} \leq K_{av} < K_{av(1-2)}$

3 $K_{av(3-4)} \leq K_{av} < K_{av(2-3)}$ 4 $K_{av} < K_{av(3-4)}$

| Predicted risk level | The actual level of SRC facility dependability | | | |
|---|---|---|---|---|
| | 1 | 2 | 3 | 4 |
| Unacceptable | 80 excellent | 75 good | 25 unsatisfactory | 20 unsatisfactory |
| Undesired | 85 excellent | 80 excellent (A) | 30 unsatisfactory | 25 unsatisfactory |
| Acceptable | 90 excellent | 85 excellent | 55 satisfactorily | 30 unsatisfactory (B |
| Negligible | 95 excellent | 90 excellent | 60 good | 55 satisfactorily |

Fig. 8.3 Quantitative and qualitative assessment of the basic quality indicator of SRC facility technical operation

| Consequence levels (loss of train hours) / Frequency levels (probability of damage) | $T_D < \frac{T_{D\,acceptable}}{K^2}$ Minor 1 | $\frac{T_{D\,acceptable}}{K^2} \le T_D < \frac{T_{D\,acceptable}}{K}$ Considerable 2 | $\frac{T_{D\,acceptable}}{K} \le T_D < T_{D\,acceptable}$ Significant 3 | $T_{D\,acceptable} \le T_D$ Critical 4 |
|---|---|---|---|---|
| $P_{D\,acceptable} \le P_D$ | acceptable | undesired | unacceptable | unacceptable |
| $\frac{P_{D\,acceptable}}{K} \le P_D < P_{D\,acceptable}$ | acceptable | undesired | undesired | unacceptable |
| $\frac{P_{D\,acceptable}}{K^2} \le P_D < \frac{P_{D\,acceptable}}{K}$ | acceptable | acceptable | undesired | unacceptable |
| $\frac{P_{D\,acceptable}}{K^3} \le P_D < \frac{P_{D\,acceptable}}{K^2}$ | negligible | acceptable | undesired | undesired |
| $\frac{P_{D\,acceptable}}{K^4} \le P_D < \frac{P_{D\,acceptable}}{K^3}$ | negligible | acceptable | acceptable | undesired |

**Fig. 8.4** Modified matrix of the predicted level of risk associated with the dependability of the SRC facility

We assume that the value of the availability coefficient of the fourth dependability level is equal to the acceptable one. The third, second, and first dependability levels are calculated according to the standard formula

$$K_{\text{av}(i-j)} = \frac{1}{1 + \lambda_{(i-j)} T_R},$$

where $T_R$ is the average time to facility recovery, and the rate $\lambda_{(i-j)}$ is the function of train delays $T_{D(i-j)}$ due to failures, pre-failures, and incidents during the operation of the SRC facility. Wherein

$$T_{D(3-4)} = T_{D\text{ acceptable}}, T_{D(2-3)} = \frac{T_{D\text{ acceptable}}}{2}, T_{D(1-2)} = T_{D\text{ acceptable}}/4.$$

The predicted risk level is estimated using the modified risk matrix given in Table 8.4. Accepted assumptions: the duration of the estimation period is 12 months; relative step along the axes of the risk matrix $K = 2$.

The acceptable probability of train delays $P_D$ associated with SRC facility dependability is defined as the ratio of the average number of failures that led to train delays to the sum of the average number of failures, pre-failures, and incidents occurred during the year. The probability of train delays is estimated by the ratio of the actual number of failures that led to train delays, to the actual number of failures, pre-failures, and incidents occurred during the year.

The basic quality indicator of the SRC facility technical operation is assessed using the predicted level of risk associated with the dependability of SRC facility, and its obtained actual level of dependability.

For example, the predicted level of risk is undesired, but the actual result is good (case A). The result of 80 scores is excellent.

**Table 8.4** Estimation scale for an additional quality indicator of technical operation of the SRC facility

| No | The share of operational failures of the SRC facility (failures caused by non-compliance with the device maintenance technology) in the total number of SRC facility failures | Assessment of indicator $A_i^F$, scores |
|---|---|---|
| 1 | Less 0.2 | 20 |
| 2 | 0.2 – 0.5 | 0 |
| 3 | More 0.5 | -20 |

**Table 8.5** Assessment scale of the general additional indicator of SI department activity $A^{SRC}$

| Range of values $A_\Sigma^{SRC}$, scores | Indicator assessment $A^{SRC}$, scores |
|---|---|
| $0.75A_{max}^{SRC} < A_\Sigma^{SRC} \leq A_{max}^{SRC}$ | 20 |
| $0.5A_{max}^{SRC} < A_\Sigma^{SRC} \leq 0.75A_{max}^{SRC}$ | 10 |
| $0.25A_{max}^{SRC} < A_\Sigma^{SRC} \leq 0.5A_{max}^{SRC}$ | 5 |
| $0 < A_\Sigma^{SRC} \leq 0.25A_{max}^{SRC}$ | 1 |
| $A_\Sigma^{SRC} = 0$ | 0 |
| $0.25A_{min}^{SRC} \leq A_\Sigma^{SRC} < 0$ | -1 |
| $0.5A_{min}^{SRC} \leq A_\Sigma^{SRC} < 0.25A_{min}^{SRC}$ | -5 |
| $0.75A_{min}^{SRC} \leq A_\Sigma^{SRC} < 0.5A_{min}^{SRC}$ | -10 |
| $A_{min}^{SRC} \leq A_\Sigma^{SRC} < 0.75A_{min}^{SRC}$ | -20 |

Another example, the predicted risk is acceptable, but the result of the actual operation is unacceptable (4 dependability level). Result is 30 scores; the assessment is unsatisfactory.

The integral quality indicator of the technical operation of the *i*th facility $I_i^F$ is calculated as the sum of the basic and additional indicators.

An additional indicator $A_i^F$ is assessed by the number of technological violations that occurred during the operation of the facility for the estimated period. Estimation scale (expressed in scores) set for an additional indicator is presented in Table 8.4.

The assessment (expressed in scores) of an additional quality indicator of the technical operation of the **SRC** facility is performed by comparing its actual value for the past period (month, quarter, and year) with the estimation scale of Table 8.5.

### *8.4.3 The General Procedure for Assessing the Integral Indicator of SI Department Activity*

As initial data, we use the quantitative values of the integral quality indicator of the technical operation of individual SRC objects within the limits of the production activity of the SI department.

The value of the basic indicator of SI department activity is calculated on the basis of the integral quality indicators of the technical operation of SRC facility within the limits of the production activity of given SI department. It is calculated according to the formula:

$$B^{\mathrm{SRC}} = \frac{\sum_{j=1}^{m} I_i^F}{m} \tag{8.20}$$

where $m$ is the number of SRC objects within the limits of the production activity of the SI department and $I_i^F$ is an integral indicator of the quality of technical operation of the $i$th SRC object.

The list of additional indicators of the SI department activities and their assessment scale expressed in scores are established by the order of the Central Infrastructure Directorate of JSC "Russian Railways" and can be revised annually. The assessment (expressed in scores) of additional indicators of the SI department activity is performed by comparing the actual values of the additional indicators of SI department activity for the past period (month, quarter, and year) with the corresponding assessment scales.

*The integral indicator of the SI department activity* $I^{\mathrm{SRC}}$ *is determined by summing up the base indicator* $B^{SRC}$ *and the general additional indicator of the SI department activity* $A^{\mathrm{SRC}}$.

The general additional indicator of SI department activity is assessed using Table 8.5.

The maximum and minimum possible sums of assessments of additional indicators are determined by the formulas:

$$A_{\max}^{\mathrm{SRC}} = \sum_{i=1}^{N} A_{i\ \max}^{\mathrm{SRC}}; A_{\min}^{\mathrm{SRC}} = \sum_{i=1}^{N} A_{i\ \min}^{\mathrm{SRC}} \tag{8.21}$$

where:

$A_{i\ \max}^{\mathrm{SRC}}$ and $A_{i\ \min}^{\mathrm{SRC}}$ are the maximum and minimum value (respectively) of the assessment of the $i$th additional indicator of SI department activity

$N$ is the number of additional indicators of SI department activity.

We determine the actual sum of additional indicators assessments with the help of the actual values (set in scores) of assessment $A_{i\ F}^{\mathrm{SRC}}$ of the ith additional indicator of the SI department activity for the reporting month (quarter and year)

$$A_{\Sigma}^{\mathrm{SRC}} = \sum_{i=1}^{N} A_{i\ F}^{\mathrm{SRC}} \tag{8.22}$$

The assessment of the general additional indicator $A^{\mathrm{SRC}}$ of SI department activity is performed with help of Table 8.5 and the use of scores. The value of general additional indicator is in right column opposite to the calculated value $A_{\Sigma}^{\mathrm{SRC}}$ that is in the same row in the left column.

## 8.5 Assessment of the Activities of Transenergo Structural Divisions

### *8.5.1 General Provisions*

As the subject of assessment, we take the results of the activities of a separate structural division of the electrification and power supply facility for a certain period that is typically a calendar year. At the same time, each activity indicator (in accordance with its purpose) is referred to a certain reference object of power supply (reference catenary system (CS) in the open track, reference CS at the station, etc.). This makes it possible to rank various structural divisions of the facility, regardless of their equipment, configuration, the number of pieces of equipment, and operating conditions.

The integral assessment characterizes the activities of Transenergo's structural divisions on the basis of a set of global (complex) indicators. Global indicators are determined on the basis of local (primary) indicators.

We use the following global indicators to obtain an integral assessment of the activity of the power supply department:

- Indicator of train traffic safety ($S_{\text{safety}}$)
- Indicator of reliability of department objects ($S_{\lambda}$)
- Indicator of failure recovery rate ($S_{\mu}$).

The assessment should be carried out within a limited period of time. Mostly, such a period is equal to a calendar year, but it can be changed (we can extend or shorten it). Hereinafter, this period will be referred to as *the assessment period*. To improve the accuracy of the assessment, we should calculate indicators for an increased time interval in comparison with assessment period. This interval will be referred to as *the calculation period*, which is recommended to be equal to three years.

The assessment of the activity of the power supply department for the calculation period should be carried out in three stages:

1. At the first stage, we assess the global safety indicator $S_{safety}$ using a two-point scale. It can take on the values "satisfactory" or "unsatisfactory." The value "unsatisfactory" is given to a structural division of the power supply department if, during the assessment period, within at least one of the power supply object maintained by this division, there was a device failure or violation of the rules for their technical operation. This led to the occurrence of transport accidents and others related to the violation of the rules of traffic safety and operation of railway transport. The safety indicator when evaluating the activity of the power supply distance acts as the next decisive rule:

   (a) If the division receives an "unsatisfactory" value according to this indicator, then its activity for the analyzed period is assessed as *unsatisfactory* regardless of the scores it receives in respect to other indicators

**Table 8.6** Quality categories of the activity of the structural division of the power supply department

| Quantitative value of the integral indicator of activity of the power supply department $S_{int}$ | Qualitative indicator of the activity of the power supply department |
|---|---|
| 0–30 | "Excellent" |
| 31–60 | "Good" |
| 61–100 | "Satisfactory" |
| >100 | "Unsatisfactory" |

(b) If the division activity in terms of the safety indicator $S_{\text{safety}}$ is assessed as "*satisfactory,*" then the value of the integral quality indicator of the structural division activity $S_{\text{int}}$ is determined by the values of the selected global indicators of the failure rate and the rate of recovery of power supply objects of section under consideration.

2. At the second stage, we assess the global indicator of reliability of department objects $S_{\lambda}$ and the rate of failures recovery $S_{\mu}$.
3. At the third stage, we determine an integral indicator $S_{\text{int}}$ that assesses the activity of the structural division of the electrification and power supply facility as a whole.

Table 8.6 gives the quantitative values of the integral indicator of the activity of the Transenergo structural division and values of the integral quality indicator associated with the quantitative values.

### 8.5.2 Determination of Particular, Generalized, and Global Indicators of Transenergo Department Activity

***A particular reliability indicator of the*** *j*th object of railway power supply facility is expressed in relative dimensionless units. The value of the particular reliability indicator of each *j*th object is calculated using the actual $\lambda_{\text{actual } j}$ and acceptable $\lambda_{\text{acceptable } j}$ values of the object failure rate in accordance with the formula:

$$S_{\lambda i} = 100\frac{\lambda_{\text{actual } j}}{\lambda_{\text{acceptable } j}} \tag{8.23}$$

The value of the particular indicator of the failure rate is less than 100 when the actual failure rate does not exceed the acceptable one. It corresponds to the range of positive categories ("satisfactory," "good," or "excellent") of the quality of the structural division activity in respect to failures of given railway power supply object.

On the basis of particular indicators of the failure rate within one section (or for all sections maintained by the structural division of the electrification and power supply facility), we calculate the ***generalized indicators of the reliability*** of department objects as a whole for railway power supply objects of each type: for CS in open lines, for CS at stations, for traction substations (TS), and for electricity transmission lines (ETL). We use the following formula to calculate the generalized indicator of the failure rate of railway power supply objects of type $z$. It takes into account the share of fitting of the section (sections) with railway power supply objects of given type.

$$S_{\lambda z} = \frac{\sum_{j=1}^{N_z} S_{\lambda zj}}{n_z} \tag{8.24}$$

where:

z = 1 is for the reference CS in open lines; z = 2 is for the reference CS at station; z = 3 is for the reference TS; z = 4 is for a reference ETL

$n_z$ is the number of railway power supply objects of type z within a given section (sections) maintained by the structural division of the power supply department.

Based on the generalized indicators calculated for the Transenergo structural division as a whole, we determine the ***global indicator of the reliability of railway power supply objects*** according to the following formula:

$$S_\lambda = \sigma \sum_{i=1}^{4} k_i S_{\lambda i}, \tag{8.25}$$

where $k_i$ is the weight coefficients for the generalized indicators of the reliability of the reference CS in open lines, the reference CS at station, the reference TS, and the reference ETL, respectively. Generally ($k1 + k2 + k3 + k4$) = 1. The values of the weight coefficients are recommended to be taken as follows: $k1 = 0.31$; $k2 = 0.28$; $k3 = 0.25$; $k4 = 0.16$. These values can be corrected for each railway taking into account local conditions; σ is the coefficient taking into account the availability of railway power supply objects of type $z$ in the power supply structural division, i.e., when ($k1 + k2 + k3 + k4$) ≠ 1:

$$\sigma = \frac{1}{\sum_{z=1}^{4} k_z y_z} \tag{8.26}$$

where $y_z$ = (1.0) depending on the availability of an object of type $z$.

We determine ***particular, generalized, and global indicators of the failure recovery rate of objects of railway power supply department*** by the following formulas that are similar to (8.23)–(8.25)

**Table 8.7** Regulatory values of the average time to recovery of the object

| Type of railway power supply object | Average time to recovery, h |
|---|---|
| Catenary system in open lines | 1.33 |
| Catenary system at station | 1.00 |
| Traction substation | 1.00 |
| Electricity transmission line | 1.33 |

$$S_{\lambda z} = 100\frac{\mu_{\text{acceptable } j}}{\mu_{\text{actual } j}} \quad S_{\mu z} = \frac{\sum_{j=1}^{N_z} S_{\mu zj}}{n_z} \quad S_{\mu} = \sigma\sum_{i=1}^{4} k_i S_{\mu i} \tag{8.27}$$

where $\mu_{\text{acceptable } j}$, $\mu_{\text{actual } j}$ are the acceptable and actual object recovery rate, respectively. The parameter $\sigma$ is determined by the formula (8.26).

The acceptable recovery rate of a railway power supply object can be calculated using the data given in Table 8.7. This Table shows the regulatory values of the average time to recovery of object. For example, the recovery rate of an object of catenary system in open lines can be defined as

$$\mu_{\text{ACSL}} = \frac{1}{1.33} = 0.75\ 1/\text{h}$$

The integral indicator for assessing the activity of the power supply department is determined using the following formula:

$$S_{\text{int}} = \{\omega_1 S_\lambda + \omega_2 S_\mu\}\ |\left(S_{\text{safety}=\text{"Satisfactory"}}\right) \tag{8.28}$$

where $\omega_1 + \omega_2 = 1$.

The quantitative values of the weight coefficients $\omega_i$ are determined by the management of Transenergo. It is advisable to assign these weight coefficients providing $\omega_1 < \omega_2$ for objects characterized by high wear and tear. In respect to new railway power supply objects, as well as to modernized or overhauled objects, there is every reason to give preference to the global indicator of reliability. In this case, weights should be specified providing the inequality $\omega_1 \geq \omega_2$.

## 8.6 Assessment of the Activity of Operational Locomotive Depots

The purpose of assessing the activities of structural divisions of the locomotive complex is to obtain an objective assessment based on the analysis of their activities on the railway network.

**Table 8.8** The list of key indicators and their values (expressed in scores) that are used to assess the activities of operational locomotive depots

| No. | Indicator | Values, scores, $A_i$ |
|---|---|---|
| 1 | Coefficient taking into account committed fires and ignitions (excluding cases occurred due to external interference in the operation of railway transport) | 25 |
| 2 | Coefficient taking into account transportation violations (due to the fault of structural divisions of the Motive-power Directorate) | 320 |
| 3 | Coefficient taking into account traffic safety violations | 35 |
| 4 | Coefficient taking into account failures of technical means | 35 |
| 5 | Coefficient taking into account train delay period due to failures | 35 |
| 6 | Coefficient taking into account train delays period due to technological violations | 35 |
| 7 | Coefficient taking into account committed accident (due to the fault of the operational locomotive depot, excluding cases that were caused by the fault of contractors and third-party organizations when performing supporting procedures) | 350 |
| 8 | Coefficient taking into account audit orders (instructions) that are not fulfilled on time | 35 |

As the subject of assessment, we take the results of the activities of operational locomotive depots for a month, quarter, and cumulative total since the beginning of the year.

We assess the activities of the structural divisions of the locomotive complex of the Russian Railways on the basis of a set of criteria that characterize the key aspects of activities of the operational locomotive depot and the regional motive-power directorate [3]. This section considers the integral assessment of the activities of the operational locomotive depot based on a set of key indicators for certain areas. Key indicators are determined on the basis of statistical and expert data and are expressed in scores (see Table 8.2).

Table 8.8 presents the list of key indicators of activities and their values (expressed in scores) that are used to assess the activities of operational locomotive depots.

The calculated actual scores of a number of key indicators (excluding those related to fire safety, occupational safety, and auditing) are multiplied by correction coefficients $k_{yi}$ that take into account the network average levels of these indicators in the previous year. Depending on the network level of these indicators for the previous year, it is conditionally accepted:

- $k_{yi}$= 1 if the level is below the network average
- $k_{yi}$= 1.5 if a level is higher than the network average, but not more than 2 times
- $k_{yi}$= 2 if the level is higher than the network average by 2 times, but less than 3 times
- $k_{yi}$= 3 if the level is higher than the network average by 3 or more times.

The ranking assessment of operational locomotive depots is carried out in three stages:

- At the first stage, the key indicators are calculated
- At the second stage, an integral assessment of the depot's activities is carried out
- At the third stage, after taking measures to eliminate the committed drawbacks and checking their effectiveness, the integral assessment is corrected.

I. *Calculation of key indicators* for assessing the activities of operational depots (taking into account the list of these indicators given in Table 8.8).

1. Calculation of scores for committed fires

$$S_{\text{fire}} = M_1 A_1 \tag{8.28}$$

where $M_1$ is the number of fires in the depot for the reporting period.

2. Calculation of scores for traffic accidents

$$S_{\text{TA}} = M_2 A_2 \tag{8.29}$$

where $M_2$ is the number of traffic accidents (namely: passages, derailments, collisions, crashes, accidents, and unauthorized movements) within the scope of the responsibility of the operational locomotive depot for 12 months.

3. Calculation of scores for traffic safety violations

$$S_{\text{TS}} = \left| \frac{R_{\text{TS}} - R_{\text{TS}}^{\Sigma}}{R_{\text{TS}}} \right| k_{y3} A_3, \tag{8.30}$$

where:

$R_{\text{TS}}$ is the ratio of the number of traffic accidents and events occurred by the fault of the operational locomotive depot (taking into account service organizations) to the linear mileage of head locomotives

$R_{\text{TS}}^{\Sigma}$ is the ratio of the total number of traffic accidents and events occurred by the fault of all operational locomotive depots (taking into account service organizations) to the linear mileage of head locomotives throughout the network.

4. Calculation of scores for violation of dependability of technical means

$$S_{\text{reliability}} = \left| \frac{R_{\text{reliability}} - R_{\text{reliability}}^{\Sigma}}{R_{\text{reliability}}} \right| k_{y4} A_4 \tag{8.31}$$

where:

$R_{\text{reliability}}$ is the ratio of the number of technical means failures of 1, 2 categories due to the fault of the operational locomotive depot (excluding service and third-party organizations) to the total locomotives mileage with respect to all types of traffic

$R^{\Sigma}_{\text{reliability}}$ is the ratio of the total number of technical means failures of 1, 2 categories due to the fault of all operational locomotive depots (excluding service and third-party organizations) to the total mileage of locomotives with respect to all types of traffic throughout the network.

5. Calculation of scores for train delays due to technical means failures

$$S_{\text{DFTM}} = \left| \frac{t_3 - t_3^{\Sigma}}{t_3} \right| k_{y5} A_5, \tag{8.32}$$

where:

$t_3$ is the total time of trains delay due to committed failures of technical means in the operational locomotive depot

$t_3^{\Sigma}$ is total time of trains delays due to technical failures throughout the network.

6. Calculation of scores for train delays due to technological violations

$$S_{\text{DTV}} = \left| \frac{t_{\text{DTV}} - t_{\text{DTV}}^{\Sigma}}{t_{\text{DTV}}} \right| k_{y6} A_6, \tag{8.33}$$

where:

$t_{\text{DTV}}$ is the number of hours of train delays caused by technological violations in the depot

$t_{\text{DTV}}^{\Sigma}$ is the total number of all hours of train delays caused by technological violations throughout the network.

7. Calculation of the scores for committed accident

$$S_{\text{accident}} = M_7 A_7, \tag{8.34}$$

where $M_7$ is the number of accident committed.

8. Calculation of the score for audit orders (instructions) not complied on time

$$S_{\text{AOI}} = M_8 A_8, \tag{8.35}$$

where $M_8$ is the number of audit orders (instructions) not complied on time.

II. *The calculation of the integral indicator* for assessing the activities of operational depots is carried out by summing the scores of the eight above-mentioned integral indicators, and also taking into account the following:

- Indicator of availability of a conformity certificate in respect to traffic safety ($K_{\mathrm{CC}}$); if there is no certificate, it is taken equal to 1; if the certificate is available, it is taken equal to 0.75
- Indicator of the results of factor analysis of operational locomotive depots ($K_{\mathrm{FA}}$); it is calculated by the situational center for monitoring and managing emergency situations of JSC Russian Railways
- Indicator of the normalized failure-free operation of the locomotive ($K_{\mathrm{FFOL}}$). It is determined using the formula:

$$K_{\mathrm{FFOL}} = \left| \frac{\widehat{\lambda}_{FL} - \widehat{\lambda}^{\Sigma}_{FL}}{\widehat{\lambda}_{FL}} \right| k_{\mathrm{Y}}, \tag{8.36}$$

where $\widehat{\lambda}_{FL}$ is the average failure rate of one locomotive in the depot; $\widehat{\lambda}^{\Sigma}_{FL}$ is the average failure rate of one locomotive over the network as a whole.

The integral indicator for assessing the activities of the operational locomotive depot is calculated using the following formula:

$$S^{\mathrm{int}} = K_{\mathrm{CC}} K_{\mathrm{FA}} K_{\mathrm{FFOL}} \left( S_{\mathrm{fire}} + S_{\mathrm{TA}} + S_{\mathrm{TS}} + S_{\mathrm{reliability}} + S_{\mathrm{DFTM}} + S_{\mathrm{DTV}} + S_{\mathrm{accident}} + S_{\mathrm{AOI}} \right) \tag{8.37}$$

Ranking (determination of a position) of the operational locomotive depot according to the integral indicator (expressed in scores) is carried out in descending order of the scores gained when assessing the depot's activities.

III. *Elimination of drawbacks and correction of the integral assessment.* If there were events related to a transport accident or an accident within structural unit for the reporting period, and the enterprise has fully took the prescribed *corrective measures*, then according to [3], the scores of the integral assessment are reduced.

We should develop *corrective measures:*

- Upon violation of traffic safety or an accident that were committed (corrective measures should be developed taking into account the measures specified in the protocol of analysis of the committed traffic safety violation)
- Upon the results of unscheduled control activities after the committed accident.

Based on the results of checking the *corrective measures*, we then determine degree (expressed in percentage) of their implementation (0%—not a single measure was completed and 100%—all measures were completed in full).

If, according to the results of the audit carried out by the auditor of the Motive-power Directorate, it is found that the *corrective measures* were completed in full (100%), then the weight coefficient corresponding to the indicator "number of transport violations" or "number of accidents committed" is reduced by 90%.

If it is determined that the amount of *corrective measures* completed is between 90% and 99%, then the weight coefficient is reduced by 80%. If it is determined that the amount of *corrective measures* completed is between 80% and 89%, then the weight coefficient is reduced by 70% and the ranking score is decreased by 10%.

The head of the operational locomotive depot is entitled to additionally reduce the ranking score if a team fully reach the norms of personal participation in ensuring traffic safety for the reporting period (from the moment of the committed traffic accident to the moment of the inspection).

If, based on the results of checking the norms that is carried out by the auditor of the Motive-power Directorate, it is found that they are reached in full (100%), then the current general ranking score is reduced by 10%.

If it is found that the norms are reached in the amount of 90–99%, then the current integral score of the ranking assessment is reduced by 7%. If it is found that the norms are reached in the amount of 80–89%, then the current integral score of the ranking assessment is reduced by 5%.

## References

1. Metodika kompleksnoj ocenki deyatel'nosti strukturnyh podrazdelenij hozyajstva avtomatiki i telemekhaniki po pokazatelyam nadezhnosti i bezopasnosti funkcionirovaniya, kachestva tekhnicheskogo obsluzhivaniya i remonta sistem i ustrojstv (Method of integrated assessment of the activities of structural units of the signalling and remote control complex in terms of dependability and safety indicators of systems and equipment functioning and quality indicators of their maintenance and repair), Moscow, RZD (2016)
2. Metodika ocenki deyatel'nosti strukturnyh podrazdelenij ekspluatacionnogo kompleksa putevogo hozyajstva OAO "RZD" po pokazatelyam nadezhnosti i bezopasnosti zheleznodorozhnogo puti (Method for assessing the activities of structural divisions of the operational complex of the JSC "Russian Railways" track facilities in the context of dependability and safety indicators of a railway track), Moscow, RZD (2018)
3. Metodika provedeniya rejtingovoj ocenki ekspluatacionnyh lokomotivnyh depo po bezopasnosti dvizheniya (Method for ranking operational locomotive depots in terms of traffic safety), Moscow, RZD (2019)

# Chapter 9
# Unified Corporate Platform URRAN (UCP URRAN)

## 9.1 General Provisions

The management of technical assets of the Russian Railways is carried out on the basis of the URRAN system which is composed of three interrelated components:

- Risk-based management methodology for the technical maintenance of railway transport facilities, the activities of structural divisions, dependability and safety of the transportation process
- Regulatory and methodological framework of the system
- Informatization of the processes of data collection and processing, management of technical assets; automation of all regulatory documents developed within the URRAN system.

The basic concepts of the URRAN methodology are described in the previous sections. Regulatory and methodological framework was developed on its basis and includes about 150 regulatory and methodological documents regulating various aspects of asset management and activities of branches. The documents cover:

- Infrastructure complex (track and structure complex, signalling and remote control complex, electrification and power supply complex, and communication complex)
- Rolling stock (locomotive, multiple unit complex, and carriage complex)
- Additional tasks in the field of fire, environmental safety, and labor protection.

Due to the vastness of calculations in respect to many objects (assets), one has been working on automation of key regulatory and methodological documents (mainly on dependability analysis, assessment of risks and activities of structural divisions, as well as on repair planning and assessment of life cycle cost). Currently, about 30% of all documents developed within the URRAN system have been automated.



I. B. Shubinsky, A. M. Zamyshlaev, *Technical Asset Management for Railway Transport*, International Series in Operations Research & Management Science 322,
https://doi.org/10.1007/978-3-030-90029-8_9

With regard to *the track and structure facilities complex and the track repair complex*, there are **24** documents being currently in force including: **4** documents regarding dependability; **3** documents regarding risk assessment; **2** documents regarding assessment of structural division activities; **6** documents regarding assessment of physical wear and tear and residual life, as well as analysis of pre-failure conditions; **9** documents regarding economic and planning aspects of repair assignment.

With regard to *signalling and remote control complex*, there are **24** documents being currently in force including: **5** documents regarding dependability; **2** documents regarding risk assessment; **2** documents regarding assessment of structural division activities; **6** documents regarding assessment of physical wear and tear and residual life, as well as analysis of pre-failure conditions; **9** documents regarding economic aspects and planning of repair assignment.

With regard to *the electrification and power supply complex*, there are **20** documents being currently in force including: **3** documents regarding dependability; **5** documents regarding risk assessment; **1** document regarding assessment of structural division activities; **9** documents regarding assessment of physical wear and tear and residual life, as well as analysis of pre-failure conditions; **2** documents regarding economic aspects and planning of repair assignment.

With regard to *the railway telecommunication complex*, there are **7** documents being currently in force including: **2** documents regarding dependability; **1** document regarding risk assessment; **1** document regarding assessment of structural division activities; **1** document regarding assessment of physical wear and tear and residual life, as well as analysis of pre-failure conditions; **2** documents regarding economic aspects and planning of repair assignment.

With regard to *the locomotive and multiple unit complex*, there are **14** documents being currently in force including: **4** documents regarding dependability; **6** documents regarding risk assessment; **4** documents regarding economic aspects and planning of repair assignment.

With regard to *the carriage complex*, there are **2** documents developed so far; they are meant to manage the dependability of facilities of this complex.

In addition, there are **7** effective regulatory documents regarding *traffic safety*; 4 effective regulatory documents regarding *labor protection and professional risk assessment*; **6** effective regulatory documents regarding *fire safety*.

Intensive automation of all regulatory documents developed within the URRAN system on the basis of the URRAN Unified Corporate Platform is being carried out.

## 9.2 Architecture, Hardware, and Software of UCP URRAN

The UCP URRAN system includes an entrance gateway, two balancer server (main and backup), main and backup servers (hosts) that host virtual application servers, as well as virtual database servers. In addition, UCP URRAN includes the main and backup synchronization servers (Fig. 9.1).

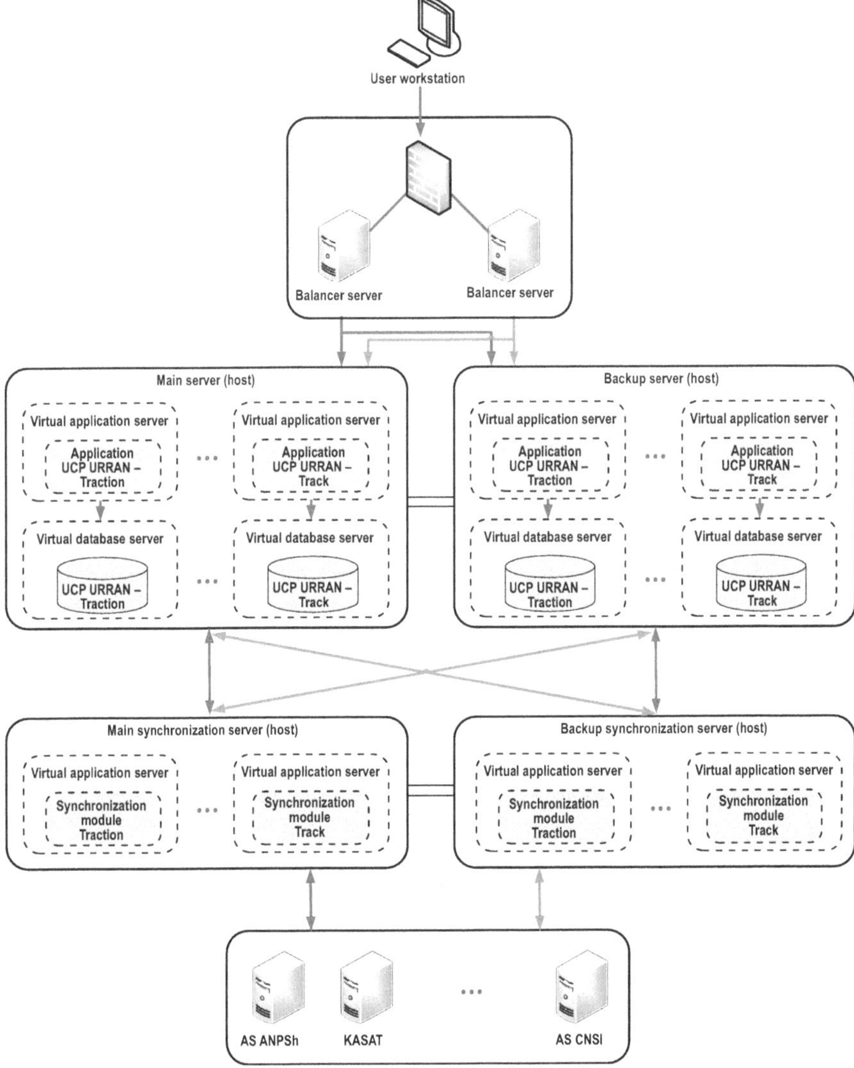


**Fig. 9.1** An enlarged diagram of the UCP URRAN architecture

The UCP URRAN system includes five functionally complete virtual technical subsystems (hereinafter Subsystems):

- UCP URRAN-Track (automation within the complex of track upper structure and structures based on the risk assessment, the processes of managing the maintenance of the track and ensuring safety during train movement, processes automation of the activity of structural divisions of the tracks and structures complex)

- UCP URRAN-Signalling (automation within the signalling and remote control complex based on risk assessment of management processes of dependability of objects of railway signalling and remote control complex, functional safety of the transportation process; management processes automation of the activities of structural divisions of the signalling and remote control complex)
- UCP URRAN-Energy (automation based on risk assessment of management processes of the technical maintenance of railway power supply and electrification facilities, fire safety; automation of assessment of activities of Transenergo structural divisions personnel and management processes of labor safety of this personnel)
- UCP URRAN-Communication (automation within the railway telecommunication complex based on risk assessment of the processes of providing telecommunication services necessary for the uninterrupted operation of railway transport (subject to the requirements of cyber security); management processes automation of the activities of structural divisions of the railway telecommunication complex)
- UCP URRAN-Traction (automation based on risk assessment of management processes of dependability and safety during the production and economic activities of locomotive and multi-unit rolling stock to ensure the possibility of maintaining an uninterrupted process of goods and passengers transportation).

Each composite Subsystem consists of the following virtual components:

- *An application server* combined with a database server;
- Balancer server
- *User* workstation
- *Administrator* workstation.

Subsystems interact through program interfaces. Fig. 9.1 gives an enlarged diagram of the UCP URRAN architecture.

To increase the reliability UCP URRAN [1] comprises four servers (Fig. 9.1). Two servers perform functions of an application server combined with a database server. The other two servers perform functions of balancers and automatically switch requests from one application server to another one if one of them fails.

If the database of one of servers fails, switching to the backup database is carried out automatically using an in-built tool (the MongoDb driver). This component is supplied with the DBMS and is responsible for receiving and transferring data from the DB to the software and makes it possible to configure in order to provide communication with several DB.

**Hardware** The UCP URRAN system is created in the form of four-layer architecture (Fig. 9.2). *Bottom layer* is data sources (automated systems KASANT, KASAT, EKASUI, ASRB, ESMA, AS CNSI, AS ANPSh, EK ASUT, etc.). *The second layer* is the integration layer that contains data integration modules. *The third layer* is data warehouses. It includes databases, aggregation functions, and compute pipeline for data aggregation. *The fourth (central) layer* is analytics layer implementing the methodology of the URRAN system.

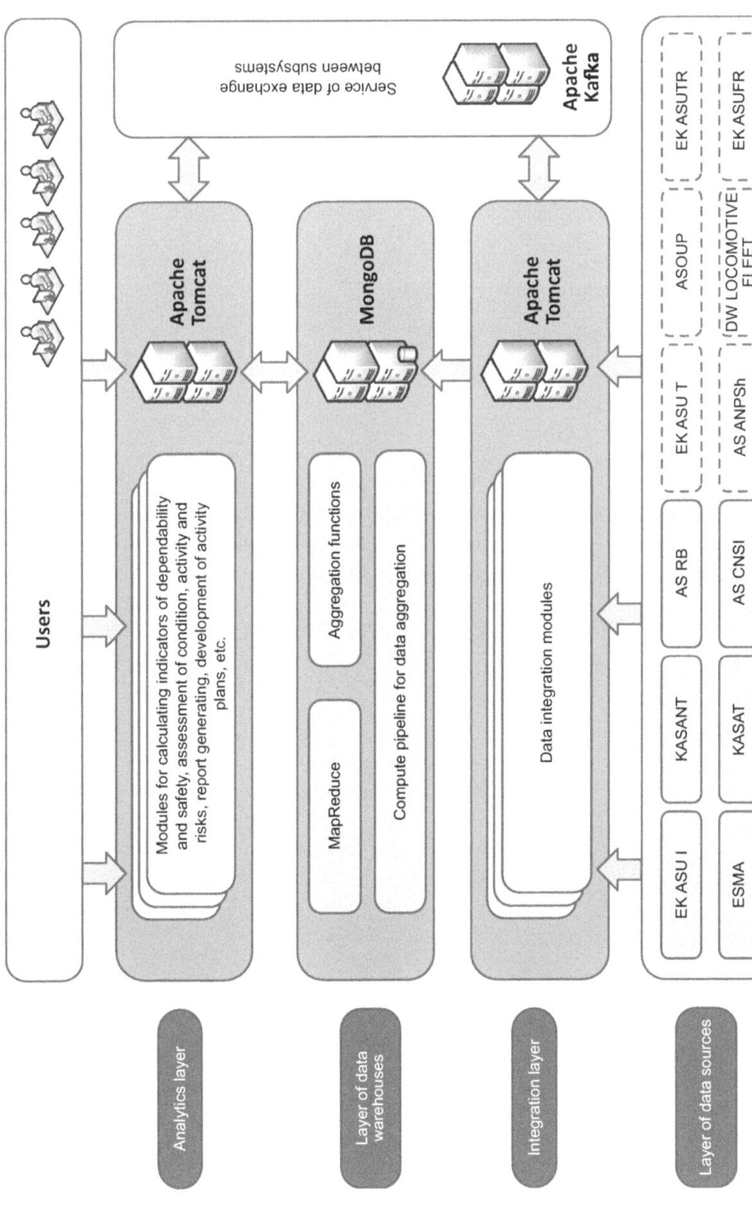


**Fig. 9.2** Four layers of UCP URRAN architecture

The UCP URRAN includes one main and one backup server (applications and databases), as well as 2 balancer servers.

Technical characteristics of each *server* (applications and databases):

- Processor: Intel Xeon 4 cores
- Ram up to 24 GB
- Disk space: up to 350 GB.

Technical characteristics of *balancer servers*:

- Processor: 2 cores Intel Xeon
- Ram: not less than 4 GB
- Disk space: up to 100 GB.

Technical characteristics of *user workstations*:

- Central processor: not less than 1.5 GHz
- RAM: more than 4 GB
- Disk space: up to 350 GB
- Network card: not less 1 Gbps
- Monitor: supports resolution not worse than 1280x720.

The UCP URRAN should provide the ability to organize application servers in a virtual environment based on the VMware platform, as well as the ability to connect users to the UCP URRAN in terminal access mode.

**Software** *System software* of the application and databases server:

- OS CentOS 7.2 or higher
- DBMS MongoDB 3.2 or higher
- Apache Tomcat 8.5 or higher.

*System software of balancer server:*

- OS CentOS 7.2 or higher
- Keepalived v1.3.5 or higher
- HaProxy 1.9.3 or higher.

*Workstation software:*

- OS Microsoft Windows 7 and higher
- Microsoft Office 2003 or higher
- Microsoft Internet Explorer 11.0 or higher, or Yandex Browser version 17.1 or higher.

UCP URRAN provides the ability to organize servers in a virtual environment based on the VMware platform and provides the ability to connect users in terminal access mode.

The application server software is implemented using client–server technologies. User's workstation and administrator's workstation are connected to the Subsystem through a local network or DTS of JSC "Russian Railways" and represent single web application (with different settings of access rights) designed to organize the interaction of the Subsystem with administrators and users.

The user WS is implemented using thin client technology (they work using Microsoft Internet Explorer 10.0 or higher) and does not require a separate installation on the user's workstation. Browser (version is not lower than 17.1) does not require a separate installation on the user's workstation.

The role model of the system provides for the following categories of users, as well as the differentiation of rights and privileges of end users:

- ***Administrators*** which include users in the role of "*Administrator*" are able to add users to the system, as well as handle all operations at all levels of the organizational hierarchy and have access to all subsystems of the system
- ***Technical administrators*** provide maintenance and support of the software and hardware complex, as well as installation of updates
- ***Technology users*** which include:
  - Users in the role of "*Editor of normative reference information (CNSI)*" can perform all operations at all levels of the organizational hierarchy and have access to all sections of the system, except for "Administration"
  - Users in the role of "Information user" are able (depending on the access level: track maintenance department, regional infrastructure directorate, or central directorate) to form parameters for calculations in all subsystems, generate and print reports, and view normative reference information
  - Users in the role of "*Technological user*" are able (depending on the access level: track maintenance department and regional infrastructure directorate) to form parameters for calculations in all subsystems, generate and print reports, view normative reference information, as well as input the necessary primary data.

## 9.3 Unified Corporate Platform URRAN-Track

The introduction of the Unified Corporate Platform URRAN to the complex of track and structures facilities (UCP URRAN Track) contributes to an overall improvement in the condition of the technical base of JSC Russian Railways, to a reduction in the number of failures of technical equipment and related damages due to the introduction of new, more dependable types of devices.

A stepwise investment in infrastructure development (in the identified key areas) is the optimum solution to maximize the return on investment. With the help of this system, rather abstract strategic goals of the Company are specified, transformed into a set of quantitative indicators and correlated to the tasks and actions of the divisions. Based on risk analysis and the availability of certain resources the system helps to monitor the Company's activities, to simulate possible scenarios of the development of the situation in the short and long term.

Table 9.1 gives the list of the UCP URRAN-Track subsystems and the list of functions they perform, in regard to which automation was carried out.

UCP URRAN Track interacts with adjacent systems:

**Table 9.1** The list of UCP URRAN-Track subsystems and the list of functions they perform in regard to which automation was carried out

| Subsystem name/functional component name | Function performed |
|---|---|
| Subsystem (as part of the UCP URRAN Track) for storing data on the objects of railway track upper structure | Storage of basic data on the objects of railway track upper structure, namely: the name of the object, location, year of commissioning; data on design and technological implementation for track section, switch; data on train traffic safety violations, failures and incidents occurred |
| Subsystem of interaction of UCP URRAN Track with external automated systems | Downloading:<br>• Guides from the AS CNSI<br>• Consolidated data on infrastructure objects from the EK ASUI<br>• Consolidated data on incidents related to objects of railway track upper structure from the EK ASUI<br>• Consolidated data on train traffic safety violations related to objects of railway track upper structure from the AS RB<br>• Consolidated data on failures from KAS ANT |
| Subsystem for maintaining reference data | Maintaining (inputting, storing, and ability of editing) information guides on a complex of track objects |
| Subsystem for collecting and processing information on pre-failure conditions and critical parameters of railway upper structure objects | Assessment of technical condition parameters in accordance with pre-failure condition classifier and reference guide on threshold values of pre-failure condition of the railway track upper structure. Formation of a summarized list of pre-failure conditions within track maintenance divisions |
| Subsystem for assessing and predicting the residual life of railway upper structure objects | Calculation of the residual life of the railway upper structure objects and formation of reporting forms on the assessment of the residual life of the railway track upper structure |
| Subsystem for reduction of existing objects of railway track upper structure to the reference object-element structure | Maintaining the reference object-element structure of the object for the linear track design, for the track crossover and crossing. Selection of polygon for calculation. Calculation of the equivalent number of reference objects at a given polygon |
| Subsystem for calculating, analyzing, and predicting dependability and safety indicators of functioning of objects of railway track upper structure | Calculation of operational dependability and safety indicators of functioning of objects of track and structure complex |

- EK ASUI industrial system (Unified corporate automated system for managing the maintenance of infrastructure) in terms of obtaining:
  - *Characteristics of infrastructure facilities*
  - *Data on malfunctions or improper maintenance of infrastructure facilities*

- KAS ANT system (Integrated automated system for recording, supervision of elimination of technical facility failures, and analysis of their dependability) *in terms of obtaining information on failures of technical means*
- AS RB system (automated system of the safety controller) in terms of obtaining *information on traffic safety violations* (*TSV*), namely:
  - Place of traffic safety violation (*TSV*)
  - Date of violation (date, hours, and minutes)
  - Cause of *TSV*
  - Damage amount caused by *TSV*
  - The division responsible for *TSV*
- AS CNSI system (Automated system of the Centralized reference data) *in terms of obtaining lists and codes of railways, lists and identifiers of interstation and open lines, and lists of organizations and enterprises.*

Subsystems interact through programming interfaces. Information exchange between subsystems is carried out through the DBMS of UCP URRAN-Track.

The UCP URRAN-Track system shall be an initiator of request on receiving data from adjacent systems and a recipient of this data. Figure 9.3 shows the scheme of information interaction of the UCP URRAN-Track.

**The module (for requesting data from the EK ASUI on the characteristics of objects of the railway track upper structure)** provides loading of summary data on infrastructure objects from the EK ASUI into the UCP URRAN Track in terms of the formation of a data set on the following objects:

- ***A track section of open lines or station*** (identifier of a railway, track maintenance department, directions, location, characteristics of track, length, class, group, category of track, type of ballast, type of rails, type of rail support, section traffic density, climatic conditions, data on latest repairs, set speeds, and technical condition)
- ***Switch*** (identifier, number, design of switch; identifier, grade of frog; project identifier and number, date and type of laying, control method, tonnage passed through, and characteristics and parameters of the track).

**Hardware and Software** UCP URRAN-Track is a virtual technical subsystem of the UCP URRAN (see item 9.2). It includes 1 main and 1 backup server (applications and databases), as well as 2 balancer servers. The UCP URRAN-Track subsystem provides the ability of organizing the UCP URRAN-Track servers in a virtual environment on the VMware platform. UCP URRAN-Track provides the ability to connect users to the UCP URRAN-Track in terminal access mode.

UCP URRAN-Track software consists of the following components:

- *An application server* combined with a database server
- *Balancer server*
- *User* workstation
- *Administrator* workstation.

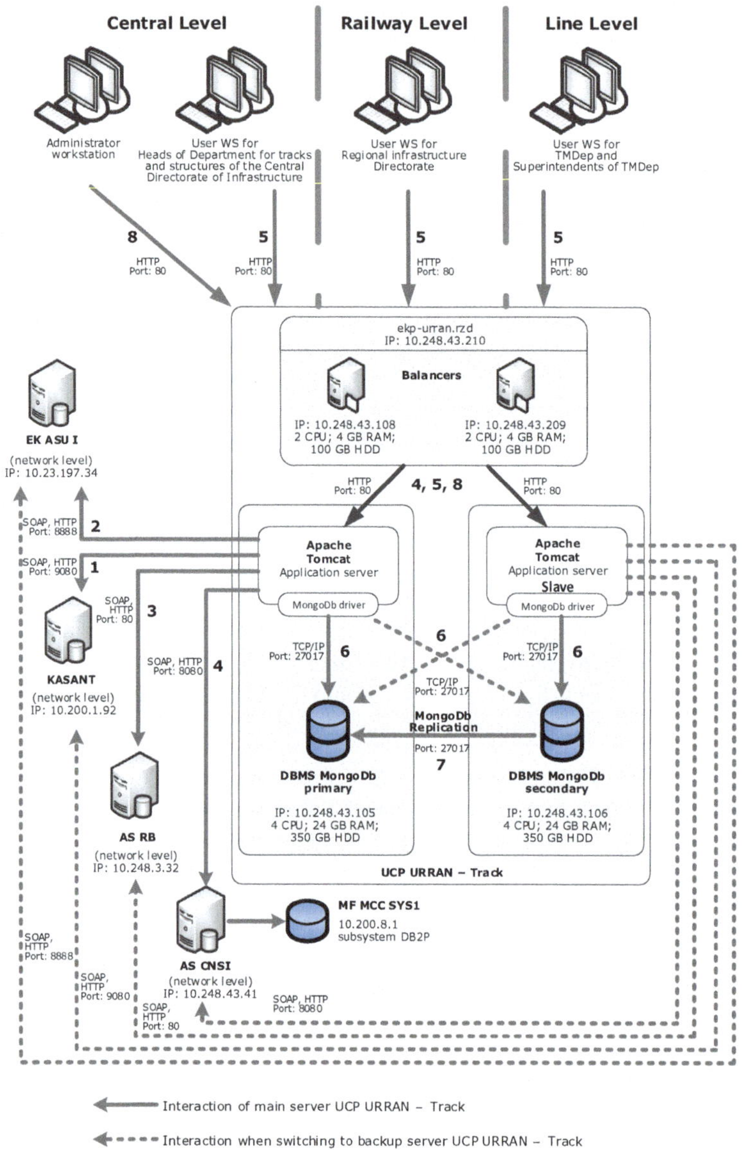


**Fig. 9.3** Scheme of information interaction of UCP URRAN Track

The application server software is implemented using client–server technologies. The user's workstation and the administrator's workstation are connected to the UCP URRAN-Track through a local network or DTS of JSC "Russian Railways." They represent one web application (with different access rights settings) designed to organize the interaction of the UCP URRAN-Track with administrators and users.

**Table 9.2** Role model of the system

| User | Role name | | | |
|---|---|---|---|---|
| | Administrator | Reference data editor | Technological user | Information users |
| Staff member of the Main Computer Center and the Technological Support Center | + | – | – | + |
| Superintendents of TMDep | – | – | – | + |
| Engineers of Engineering sector of TMDep | – | + | + | + |
| Heads of Territorial Directorates of Infrastructure | – | – | – | + |
| Engineers of Engineering sector of Territorial Directorates of Infrastructure | – | + | + | + |
| Heads of Central Directorates of Infrastructure | – | – | – | + |
| Engineers of Engineering sector of Central Directorates of Infrastructure | – | – | – | + |

The users of the UCP URRAN-Track are:

- Administrators of the Main Computer Center and technologists of the Technological Support Centers
- Superintendents and specialists of track maintenance departments
- Heads and specialists of territorial directorates of infrastructure
- Heads and specialists of the Department for tracks and structures of the Central Directorate of Infrastructure.

The role model of UCP URRAN-Track is shown in Table 9.2

## 9.4 Unified Corporate Platform URRAN-Signalling

The purpose of creating the UCP URRAN-Signalling system is to increase the efficiency and safety of production and economic activities of the complex of signalling and remote control facilities, improve the condition of the technical facilities, and reduce the number of failures of technical means.

The UCP URRAN-Signalling system is used to automate the activities of:

- Signalling and remote control department of the Central Directorate of Infrastructure—the Russian Railways branch (itself)
- Signalling and remote control division of Regional Infrastructure Directorates
- Signalling and remote control department of the Infrastructure Directorate.

Within the framework of application of the UCP URRAN-Signalling in the complex of signalling and remote control facilities, the following practical tasks are solved:

- Formation and maintenance of a unified database of the calculation results of indicators of dependability and functional safety for a comprehensive state assessment of the infrastructure of the SRCF (from various adjacent systems)
- Assessment of risks associated with the dependability and functional safety of SRCF
- Assessment and management of the life cycle cost of railway signalling and remote control systems
- Assessment and analysis of the influence of the human factor on indicators of dependability and safety of the functioning of railway signalling and remote control systems.

Table 9.3 gives the list of the UCP URRAN-Signalling subsystems and the list of functions they perform in regard to which automation was carried out.

There is an individual UCP URRAN-Signalling subsystem to solve each of the listed tasks.

The UCP URRAN-Signalling system interacts with the following adjacent systems:

- KAS ANT—in terms of obtaining data on failures of SRC technical means
- KASAT (integrated automated systems for recording, investigation, and analysis of technological violations)—*in terms of obtaining data on technological violations*
- AS RB—in terms *of obtaining information on traffic safety violations (TSV)*
- AS ANPSh (Automated system for statistical analysis of indicators of dependability and prescriptive processes management of a complex of signalling and remote control) *in terms of obtaining*:
  - Actual and normalized indicators of dependability and safety of functioning of SRCF, taking into account the railway line class and specialization
  - Performance (type of signalling and remote control system, number of points/block sections, and regulatory time for failure elimination)
  - Data on the risk assessment performed in respect to SRCF
  - Basic, additional, and integral indicators of assessment of structural division activities
- AS CNSI *in terms of obtaining information on the general characteristics* of SRCF (name, location, line class, and specialization) as well as data on design and technological implementation.

Sources and consumers of information and exchange protocols are given in Table 9.4

**Hardware and Software** UCP URRAN-Signalling is a virtual technical subsystem of UCP URRAN (see item 9.2). It includes 1 main and 1 backup server (applications

**Table 9.3** The list of functions in regard to which automation in the UCP URRAN was carried out

| Subsystem name/functional component name | Function performed |
|---|---|
| A unified database of calculation results of indicators of dependability and functional safety for a comprehensive state assessment of the infrastructure of the SRCF (from various adjacent systems) | Information interaction with KAS ANT in terms of implementation of the functionality of obtaining data on SRC technical means failures.<br>Information interaction with KASAT in terms of the implementation of the functionality of obtaining data on technological violations.<br>Information interaction with the AS RB in terms of the implementation of the functionality of obtaining information on TSV.<br>Information interaction with AS CNSI in terms of obtaining data from industrial reference data and classifiers.<br>Information interaction with the AS ANPSh in terms of obtaining:<br>• Actual and normalized indicators of dependability and safety of functioning of SRCF, taking into account the railway line class and specialization<br>• General characteristics of SRCF (such as name, location, line class, and specialization), as well as data on the design and technological implementation (type of signalling and remote control system, number of points/block sections, and regulatory time for failure elimination)<br>• Data on the risk assessment performed in respect to SRCF<br>• Basic, additional, and integral indicators of assessment of structural division activities.<br>Displaying received data. |
| Subsystem for assessing risks associated with the dependability and functional safety of SRCF | Displaying the risk matrix for the given conditions.<br>Displaying the ranking of sections in accordance with the risk assessment performed in order to make a decision when managing technical maintenance. |
| Subsystem for assessing the life cycle cost of railway signalling and remote control systems | Input of data on costs associated with research, development, and investment.<br>Determination of costs associated with dependability.<br>Determination of costs associated with disposal.<br>Displaying the LLC estimation obtained. |

and databases), as well as 2 balancer servers. The UCP URRAN-Signalling subsystem provides the ability of organizing the UCP URRAN-Signalling servers in a virtual environment on the VMware platform. UCP URRAN-Signalling

**Table 9.4** Sources and consumers of information and exchange protocols

| Source of information | Consumer of information | Request initiator | Output data | Exchange protocol |
|---|---|---|---|---|
| KAS ANT | UCP URRAN-Signalling | UCP URRAN-Signalling | Place, time, cause, object, failure mode; structural division, number of trains delayed, and loss of train-hours | HTTP // XML |
| KASAT | UCP URRAN-Signalling | UCP URRAN-Signalling | Place, time, cause, object, structural division, mode of the technological violation; number of trains delayed and loss of train-hours | HTTP // XML |
| AS RB | UCP URRAN-Signalling | UCP URRAN-Signalling | Place, date, cause of traffic safety violations (TSV); amount of damage caused by TSV and structural division | HTTP // XML |
| AS CNSI | UCP URRAN-Signalling | UCP URRAN-Signalling | List of signalling departments and superintendents of signalling departments | HTTP // XML |
| AS ANPSh | UCP URRAN-Signalling | UCP URRAN-Signalling | 1. Normalized, acceptable, and actual values of dependability and safety indicators, taking into account the class and specialization of railway lines, and the acceptable value of train delay probability.<br>2. General characteristics of SRCF.<br>3. Data on risk assessment performed: railway, structural division, SRCF, risk probability, amount of damage, observation time, and risk level.<br>4. Assessment of structural division activities | JSON // XML |

provides the ability to connect users to UCP URRAN-Signalling in terminal access mode.

UCP URRAN-Signalling software consists of the following components:

- An application server combined with a database server
- *Balancer server*
- *User* workstation
- *Administrator* workstation.

The application server software is implemented using client–server technologies. The user's workstation and the administrator's workstation are connected to the UCP URRAN-Signalling through a local network or DTS of JSC "Russian Railways." They represent one web application (with different access rights settings) designed to organize the interaction of the UCP URRAN-Signalling with administrators and users.

Subsystems interact through programming interfaces. Information exchange between subsystems is carried out through the DBMS of UCP URRAN-Signalling.

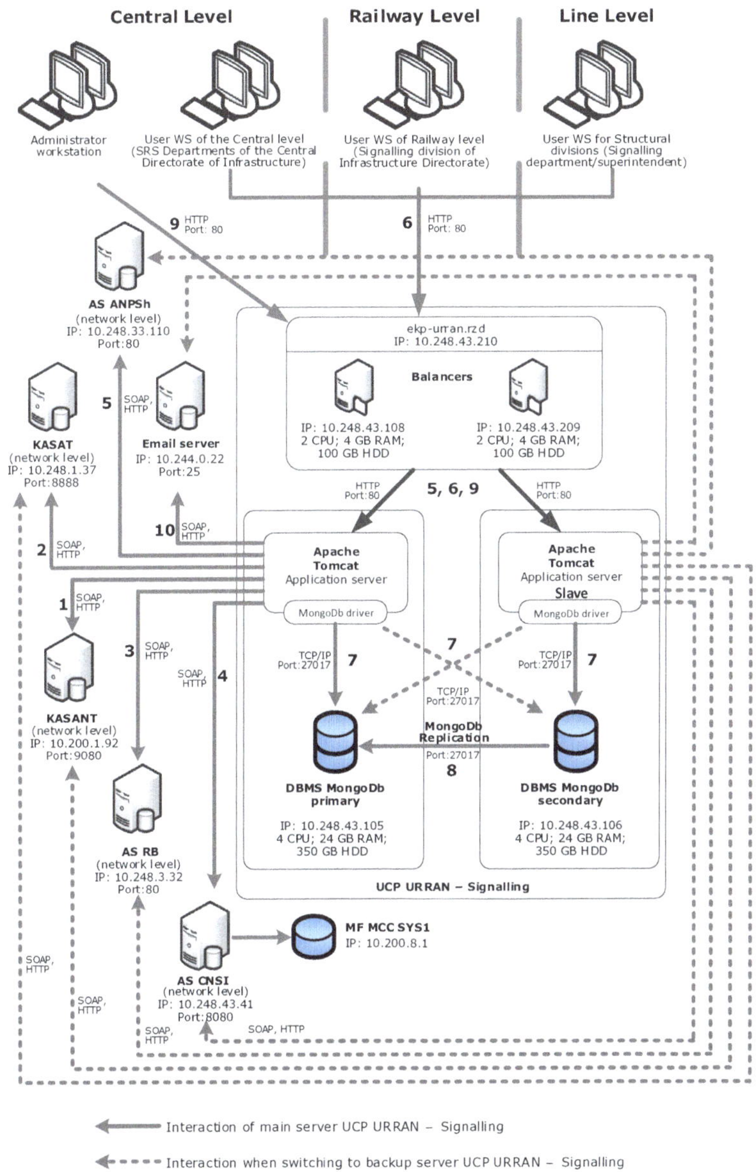


**Fig. 9.4** Scheme of information interaction of UCP URRAN-Signalling

**Table 9.5** Role model of UCP URRAN-Signalling

| User | Role name | | | |
|---|---|---|---|---|
| | Administrator | Reference data editor | Technological user | Information user |
| Staff member of the Main Computer Center and the Technological Support Center | + | – | – | + |
| Heads of Signalling departments | – | – | – | + |
| Heads of Signalling divisions | – | –– | – | + |
| Engineers of the Technical sector of Signalling division | – | + | + | + |
| Heads of SRC Departments of the Central Directorate of Infrastructure. | – | – | – | + |
| Engineers of Technical sector of SRC Departments of the Central Directorate of Infrastructure | – | + | – | + |
| Specialists of other divisions | – | – | – | + |

Figure 9.4 shows the scheme of information interaction of UCP URRAN-Signalling.

The users of the UCP URRAN-Signalling are:

- Administrators of the Main Computer Center and technologists of the Technological Support Centers
- Superintendents and specialists of signalling departments
- Heads and specialists of territorial directorates for infrastructure
- Heads and specialists of the SRC Department of the Central Directorate of Infrastructure.

The role model of UCP URRAN-Signalling is shown in Table 9.5.

## 9.5 Unified Corporate Platform URRAN-Energy

UCP URRAN-Energy subsystem is designed for:

- Improving the condition of the technical base of JSC "Russian Railways" by: calculating and analyzing indicators of operational dependability and functional safety for railway power supply objects
- Analysis and comparison of various scenarios to support management decisions based on reliable results of the analysis of dependability indicators
- Risk assessment for railway power supply objects
- Decision support when drawing up plans for repairs and reconstruction/modernization of railway power supply objects

- Increasing the efficiency and safety of production and economic activities of JSC "Russian Railways"
- Analyzing labor safety indicators in power supply departments and assessing occupational risks
- Assessing the activities of structural divisions for railway power supply.

The UCP URRAN subsystem is used to automate activities of:

- Transenergo—JSC "Russian Railways" branch
- Directorates for energy supply
- Power supply departments.

Table 9.6 gives the list of the functional components of the UCP URRAN-Energy and the list of functions performed by them in regard to which the automation was carried out.

The UCP URRAN-Energy system interacts with adjacent systems:

- EK ASUI industrial system *in terms of obtaining characteristics of infrastructure objects and data on malfunctions or improper maintenance of infrastructure objects*
- KAS ANT *in terms of obtaining data on failures of technical means of SRCF*
- AS RB *in terms of obtaining information on TSV*
- AS CNSI.

Figure 9.5 shows the scheme of information interaction of the UCP URRAN-Energy.

The users of the UCP URRAN-Energy are:

- Administrators of the Main Computer Center and technologists of the Technological Support Centers
- Personnel of line divisions (superintendent of sector for catenary system, head of traction substation, superintendent of sector for power supply, repair and revision sector)
- Superintendents and specialists of power supply department
- Heads and specialists of territorial directorates for energy supply
- Heads and specialists of Transenergo—JSC "Russian Railways" branch.

The role model of the system is shown in Table 9.7.

## 9.6 Unified Corporate Platform URRAN-Communication

The efficiency of the functioning of railway transport is largely determined by the dependability degree of the functioning of railway telecommunication facilities. The tasks in terms of ensuring the transportation process solved by the complex of railway communication objects are to provide telecommunication services necessary for the uninterrupted operation of railway transport in compliance with cyber

**Table 9.6** List of functions in regard to which the automation was performed in the UCP URRAN-Energy

| Subsystem name/functional component name | Function performed |
|---|---|
| Subsystem for collecting information on a complex of electrification and power supply objects | Collection of information on: types and categories of assets, results of inspections, tests and diagnostics of equipment, failures of technical means, and violations of traffic safety |
| Subsystem for reduction of existing railway power supply objects to the reference object-element structure | Maintaining the reference object-element structure: tension section of the catenary line in open lines and at stations, the traction (transformer) substation, and the section of the electrical transmission line.<br>Calculation of the equivalent number of reference objects at a given polygon with the reduction of this object to the reference one using conversion coefficients |
| Subsystem for assessing physical wear and residual life | Processing of the results of filling in the score cards (SC) of the states of railway power supply objects. Estimation of service and residual life |
| Subsystem for calculating and analyzing indicators of dependability and safety of functioning of railway power supply objects | Calculation of actual and acceptable indicators of dependability and safety. Formation of a report on dependability indicators of technical means at department, directorate levels, and Transenergo level in general |
| Subsystem for assessing risks associated with the functioning of railway power supply objects | Determination of the frequency of occurrence (or intensity) of dangerous failures of railway power supply objects.<br>Determination of the amount of specific damage (consequences) per failure. Determination of the risk level for a section as a whole |
| Subsystem for assessing the activities of structural divisions of the electrification and power supply facility | Determination of an integral indicator that evaluates the activity of a structural division of the electrification and power supply facility as a whole.<br>Conducting a ranking assessment of structural division with the construction of a ranking table |
| Subsystem for assessing the cost of life cycle of objects | Recording data necessary for calculating the cost of life cycle of an object in terms of the work carried out as part of the renovation and reconstruction, overhaul and maintenance |
| Subsystem for drawing up plans for the repair and renovation of railway power supply objects | Formation of a decision-making matrix.<br>Formation of a list of objects that should be included in the overhaul, modernization, or reconstruction plan |
| Subsystem for assessing occupational risks | Analysis of labor safety indicators in power supply department. Assessment of occupational risks at the central level, at the regional level, and in power supply departments.<br>Drawing up plans for improving labor |

(continued)

**Table 9.6** (continued)

| Subsystem name/functional component name | Function performed |
|---|---|
| | protection conditions for employees of power supply department |
| Subsystem for assessing fire risks of traction substations | Performing calculations to assess the probability of occurrence of fire hazardous events and the consequences of a fire at a traction substation; performing calculation of the personnel individual fire risk in a building and on the territory of traction substations |

security requirements. The use of UCP URRAN-Communication in the railway telecommunication complex makes it possible to increase the efficiency and safety of production and economic activities of the structural divisions of the complex, increase the dependability and uninterruptedness of operation of the communication network equipment.

Availability of up-to-date information on the actual state of railway telecommunication objects allows us to determine and minimize the risks that occur when managing the technical maintenance of these objects at various stages of their life cycle, and to increase the promptness and quality of decisions made.

Real-time information on the actual state of objects ensures uninterrupted railway communication, the objectivity of the assessment of the residual technical and functional life of telecommunication objects.

The UCP URRAN-Communication subsystem provides for the calculation of dependability indicators and risk assessment for the following types of communication networks:

- Communication lines (CCL, FOCL, and ACL)
- Primary network (digital and analog equipment)
- Secondary network (digital and analog equipment)
- Technological radio communication (digital and analog equipment).

Table 9.8 gives the list of the functional components of UCP URRAN-Communication and the list of functions they perform in respect to which the automation was carried out.

The interaction of the UCP URRAN-Communication with adjacent systems should be carried out at the network level through the DTS channels of JSC "Russian Railways" (see Fig. 9.6):

- KAS ANT in terms of *obtaining data on failures of technical means of railway telecommunications* and *losses caused by failures of technical means (train-hours)*;
- ESMA (Unified System for Monitoring and Administration of the Communication Network) in terms of:

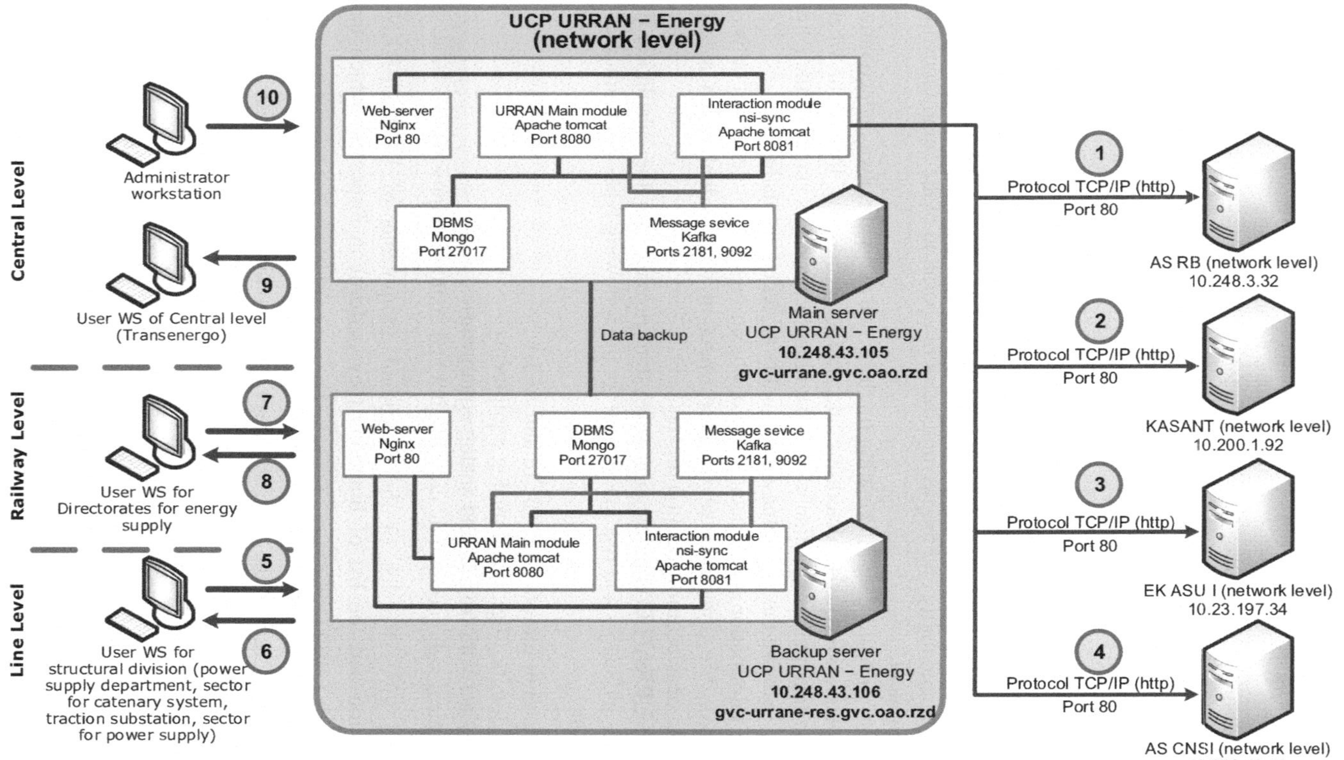


**Fig. 9.5** Scheme of information interaction of UCP URRAN-Energy

**Table 9.7** Role model of the system

| User | Role name | | | |
|---|---|---|---|---|
| | Administrator | Reference data editor | Technological user | Information user |
| Staff member of the Main Computer Center and the Technological Support Center | + | – | – | + |
| Specialists of repair and revision sector, sector for catenary system sector for power supply, and traction substation | – | – | + | + |
| Superintendents of power supply department | – | – | – | + |
| Engineers of engineering sector of power supply department | – | + | + | + |
| Labor protection specialist of power supply department | – | + | + | + |
| Heads of territorial directorates for energy supply | – | – | – | + |
| Engineers of engineering sector of territorial directorates for energy supply | – | + | + | + |
| Labor protection specialists of energy supply directorates | – | + | + | + |
| Engineers of railway electrotechnical laboratory | – | + | + | + |
| Management of Transenergo | – | – | – | + |
| Engineers of engineering sector of Transenergo | – | + | – | + |
| Labor protection specialist and engineers of department for labor protection, industrial safety, and environmental control of Transenergo | – | + | – | + |
| Specialists of other divisions | – | – | – | + |

- **Obtaining data**: *on railway telecommunication devices which failures were recorded in the KAS ANT; on failures and pre-failures with reference to the communication network and the division; on technical maintenance*
- ***Transmitting data:*** *on the normalized indicators of the dependability of railway telecommunication technical means; on the risks (risk matrix) associated with the functioning of railway telecommunication objects; on the integral assessment of the activities of structural divisions*

AS CNSI in terms of obtaining data *from industrial-wide reference books and classifiers.*

The scheme of information interaction between UCP URRAN-Communication and ESMA is shown in Fig. 9.6.

**Table 9.8** List of automated functions in the UCP URRAN-Communication

| Subsystem name | Performed function |
|---|---|
| Subsystem for interaction with external automated systems | Receiving (from KAS ANT) data on failures of technical means of railway telecommunication.<br>Receiving (from ESMA) data on railway telecommunication devices which failures were recorded in KAS ANT.<br>Transfer of following data from UCP URRAN-Communication to ESMA: data on normalized indicators of dependability of technical means, data on risks (risk matrices) and integral assessment of the activities of structural divisions.<br>Interaction with AS CNSI in terms of obtaining data from industrial-wide reference books and classifiers |
| Subsystem for calculating and analyzing indicators of dependability and safety of functioning of railway telecommunication objects | Calculation of the actual and normalized dependability indicators in respect to the communication facility for various network configurations. Comparison of the actual calculation results of the dependability indicators with the set acceptable dependability indicators |
| Subsystem for assessing risks associated with the functioning of railway telecommunication objects | Determination of the frequency of occurrence (or intensity) of each undesired event. Determination of the amount of damage (consequences) per one failure. Building a risk matrix for given conditions. Ranking of networks sections according to the performed risk assessment to make decisions when managing technical maintenance |
| Subsystem for assessing the activities of structural divisions of the communication facility | Calculation of the integral assessment of the activities of structural divisions using a set of score assessments of activities. The score assessments, in turn, are determined based on primary dependability indicators.<br>Ranking assessment of activities of communication directorates and regional communication centers.<br>Development of measures to improve the efficiency of division activities |
| Subsystem for storing data on railway telecommunication objects as part of UCP URRAN-Communication | Storage of basic data on railway telecommunication objects, consolidated data on incidents associated with railway telecommunication objects, data on its failures |
| Subsystem for maintaining normative and reference information as part of UCP URRAN-Communication | Maintaining (inputting, storing, and editing) reference books on the communication facilities (these reference books are necessary for the well functioning of the System) |
| Administration subsystem | Maintaining the registry of System users (registering and changing user credentials). Logging the activity of using communication objects |

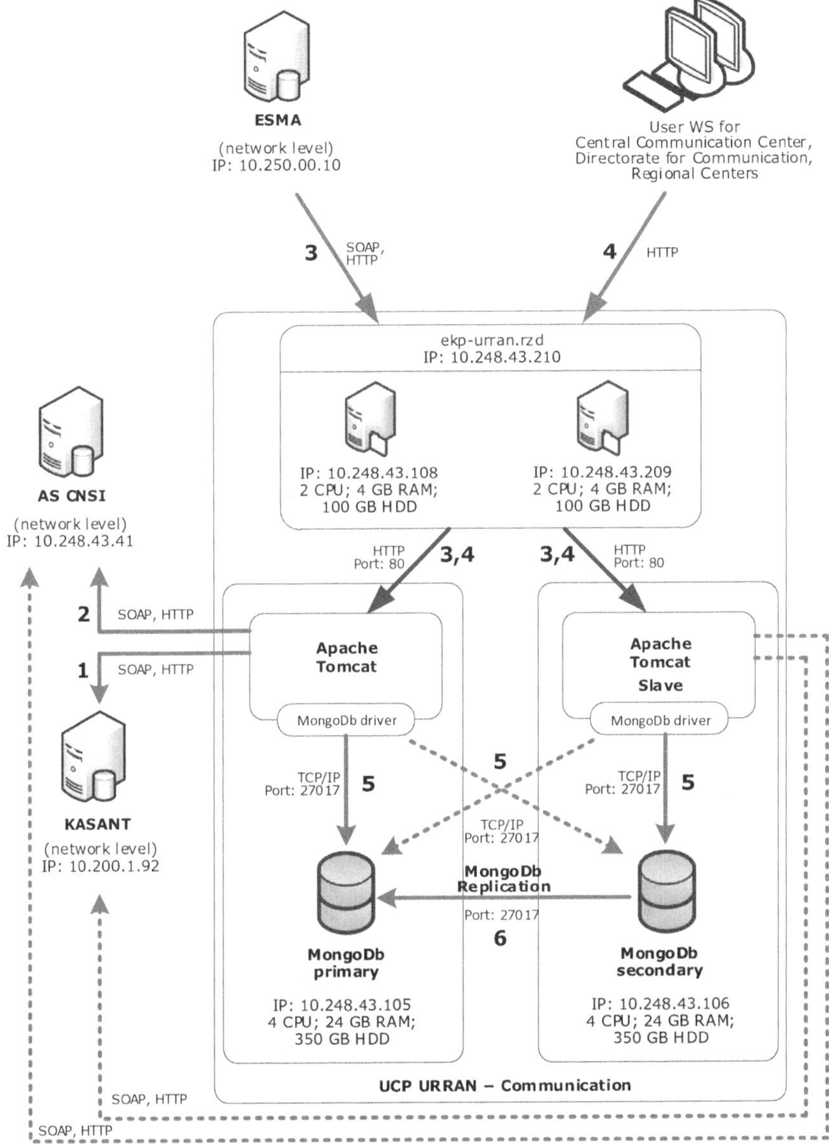


**Fig. 9.6** Scheme of information interaction of UCP URRAN-Communication

## Module for Receiving Data from ESMA

A mechanism for obtaining (from ESMA to UCP URRAN-Communication) data on railway telecommunication devices was implemented. These data include: ***data on equipment*** (name, type, category of network, date of commissioning, structural division, type of communication network and length of communication lines, number of nodes and sections of communication lines, etc.). Data upload from

ESMA to UCP URRAN-Communication is carried out once a month; ***data on failures and pre-failures*** (date and time, duration of failures and pre-failures, structural division, type of communication network, category "Service failure," "Object failure," or "Pre-failure"). Data upload from ESMA to UCP URRAN-Communication is carried out 2 times a month; ***data on technical maintenance*** (identifier from incident record sheet of the technological process chart, equipment identifier, date, type, and duration of technical maintenance, structural division, and type of communication network). Technical maintenance data upload from ESMA to UCP URRAN-Communication is carried out once a day.

**Module for Data Transmission to ESMA from UCP URRAN-Communication**
A mechanism for transferring calculated data on railway telecommunication objects from UCP URRAN-Communication to ESMA was implemented. These data include: ***normalized values of dependability indicators*** (failure rate, average time to recovery, and availability coefficient (excluding pre-failures and taking them into account)); ***actual dependability levels*** (failure rate, average and total time to recovery, and availability coefficient (excluding pre-failures and taking them into account), number of failures of 1–2 categories and losses caused by them and expressed in train-hour, increase or decrease of these indicators expressed in percent); ***design values of dependability indicators*** (design coefficients of availability (excluding pre-failures and taking them into account)); ***assessments of risks*** associated with functioning of railway telecommunication objects; ***assessments of activities of structural divisions*** of the complex of communication objects. Data upload from UCP URRAN-Communication to ESMA is carried out 2 times a month.

The scheme of information interaction of the UCP URRAN-Communication and ESMA is shown in Fig. 9.7.

The users of the UCP URRAN-Communication subsystem are:

- Administrators of the Main Computer Center and technologists of the Technological Support Centers
- Heads and specialists of Regional Communication Centers
- Heads and specialists of the Directorate for Communication
- Heads and specialists of the Central Communication Center.

The role model of the system provides for the following categories of users, as well as the differentiation of the rights and privileges of end users (Table 9.9).

## 9.7 Unified Corporate Platform URRAN-Traction

The purpose of creating "Unified corporate platform for managing resources, risks and dependability at the stages of the life cycle in the locomotive complex (UCP URRAN-Traction)" is to increase the dependability, safety of production and economic activities of locomotive complex and MU rolling stock complex in order to make it possible to maintain an uninterrupted process of transportation of goods and passengers.

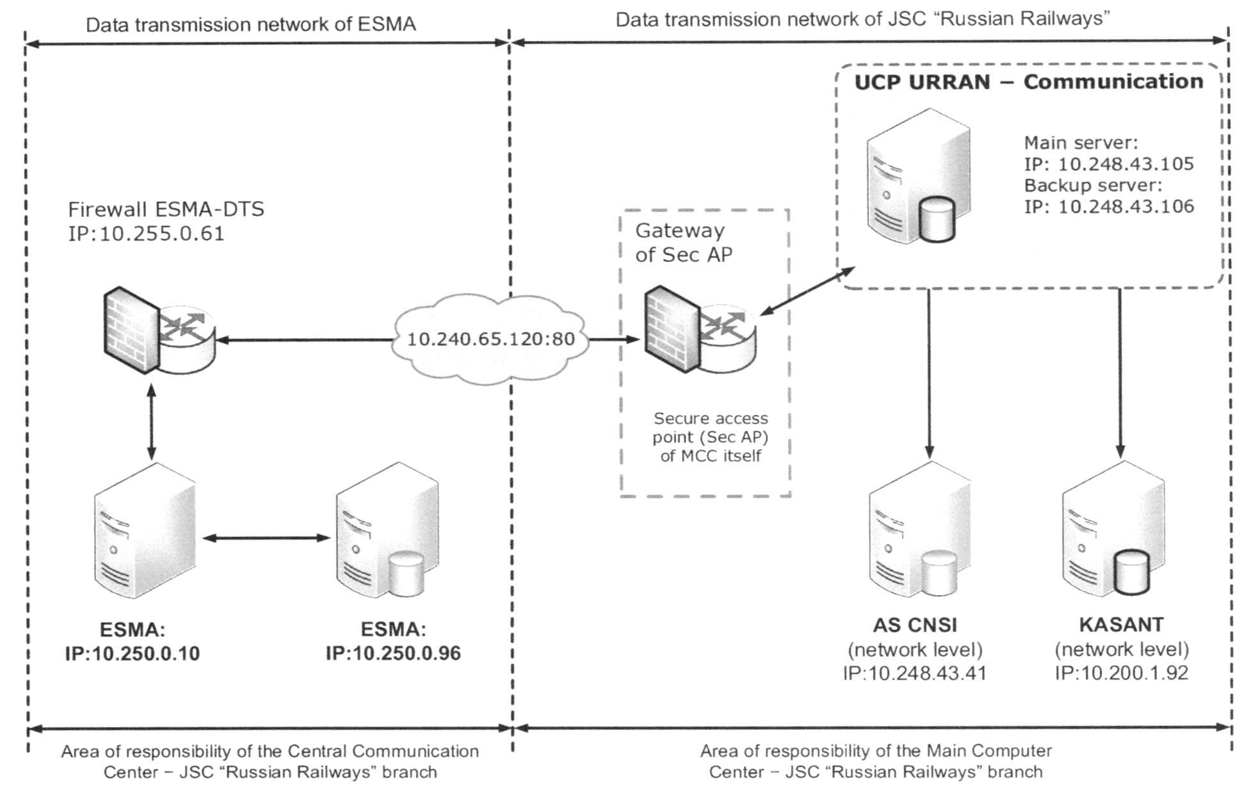


**Fig. 9.7** Scheme of information interaction of UCP URRAN-Communication and ESMA

**Table 9.9** Role model of the UCP URRAN-Communication subsystem

| User | Role name | | |
|---|---|---|---|
| | Administrator | Editor of normative reference information | Technological user |
| Staff member of the Main Computer Center and the Central Communication Station | + | – | + |
| Heads of the Central Communication Station | – | – | + |
| Specialists of the Central Communication Station | – | + | + |
| Heads of the Directorate for Communication | – | – | + |
| Specialists of the Directorate for Communication | – | + | + |
| Heads of the Regional Communication Centers | – | – | + |
| Specialists of the Regional Communication Centers | – | – | + |

The URRAN methodology makes it possible to analyze the efficiency of locomotive functioning with different levels of detail of such analysis, perform an assessment at the operational stage and make a decision on the advisability of banning locomotive operation, assess the quality of maintenance of locomotives at various levels of the locomotive facility (operational depot, railway, and railway network in general), and plan the operating costs of locomotive maintenance in order to minimize the cost of the life cycle and solve other tasks.

Previously, calculations were performed manually, which significantly increased the load on specialists. Automation has significantly reduced the burden on specialists and increased the speed of obtaining the information necessary for decision-making.

The UCP URRAN-Traction system is designed to automate the following processes:

- Calculation and analysis of indicators of operational dependability and functional safety of locomotives
- Assessment of operational risks associated with the operation of locomotives
- Assessment of fire risks associated with the operation of locomotives
- Assessment of the activities of structural divisions.

UCP URRAN-Traction is used to automate activities of:

- Traction Directorate—JSC "Russian Railways" branch (itself)
- Regional Traction Directorates
- Operational Locomotive Depots.

The list of automated functions in the UCP URRAN-Traction system is given in Table. 9.10.

**Table 9.10** List of functions automated in UCP URRAN-Traction

| Subsystem name/functional component name | Performed function |
|---|---|
| Subsystem for interaction with external automated systems | Service of information interaction with KAS ANT, KASAT, AS RB, data warehouse "Locomotive fleet," EK ASUT, ASOUP, EK ASUTR, and AS CNSI.<br>Displaying the obtained data |
| Subsystem for calculating and analyzing operational dependability and functional safety of locomotives | Calculation of indicators of dependability and safety of the locomotive depot, the coefficient of efficiency of the repair organization, and the internal availability coefficient.<br>Inputting data on the planned values of the failure number at the central level of management |
| Subsystem for assessing operational risks associated with the functioning of locomotives | Maintaining classifiers of types of traffic safety violations.<br>Calculation of the intensity and probability of traffic safety violations.<br>Formation of a register of risks of traffic safety violations.<br>Normalization of risks in the field of functional safety of traffic.<br>Calculation of operational risks and displaying operational risks calculated. |
| Subsystem for assessing fire risks associated with the functioning of locomotives | Maintaining classifiers of fire hazardous conditions of locomotives.<br>Input of data on the number of fires with a breakdown into series of locomotives<br>Implementation of forms of inputting data into electronic forms of score cards of the actual condition of locomotives.<br>An individual assessment of the condition of the locomotive, creation of a register of fire risks of traction rolling stock at the level of the operational locomotive depot, and displaying information on the identified and eliminated fire hazardous condition of the locomotives of the operational locomotive depot.<br>An integral assessment of the fire risk of locomotive series at the regional and central management levels, and a statistical assessment of the fire risk level of locomotive series at the central management level |
| Subsystem for assessing the activities of structural divisions of the locomotive facility | Maintaining the classifiers and calculating key activity indicators, calculating the overall score of an operational locomotive depot, and displaying a report on ranking of operational depots |

Data exchange between the components of the System is carried out via the TCP/IP, HTTP, and SOAP protocols within the DTS of JSC "Russian Railways."

The UCP URRAN-Traction system interacts with the following adjacent systems (Table 9.11):

**Table 9.11** Sources and consumers of information and exchange protocols in the UCP URRAN-Traction

| Source of information | Request initiator | Input information | Output information | Interaction type | Periodicity | Exchange protocol |
|---|---|---|---|---|---|---|
| KAS ANT | UCP URRAN-Traction | A period of time | Place, time, cause, mode of failure; structural division; number of trains delayed; loss of train-hours | Web service | Once a day | HTTP // XML |
| KASAT | UCP URRAN-Traction | A period of time | Place, time, cause, mode of the technological violation; structural division; number of trains delayed; loss of train-hours | Web service | Once a day | HTTP // XML |
| AS RB | UCP URRAN-Traction | A period of time | Place, time, cause of traffic safety violation (TSV), the amount of damage caused by TSV; structural division. | Web service | Once a day | HTTP // XML |
| AS CNSI | UCP URRAN-Traction | A period of time | Enterprises and organizations | Web service | Once a day | HTTP // XML |
| DW "Locomotive fleet" | UCP URRAN-Traction | A period of time | The volume of work performed by the traction rolling stock (TRS) units; total mileage of the TRS unit in the context of all types of traffic; linear mileage of head locomotives over the network. | SP | Monthly | SESSION table |
| DB UCP URRAN-Traction | User | Filter options | Creation of reporting forms | Web interface | By user request | HTTP |
| Applications server | UCP URRAN-Traction | Data from 1, 2, 3, 4, 5, 6, and 9 | Data required to create reporting forms upon request 6. | Functions | By user request | |
| Main DB UCP URRAN-Traction | UCP URRAN-Traction | Data from 1, 2, 3, 4, 5, 6, and 9 | Formation of a copy of data | Email | Scheduled | - |
| | Administrator | User data | Maintenance of user register | | | - |

| Administrator's workstation | | | | Web interface | By user request | |
|---|---|---|---|---|---|---|
| Administrator's workstation | Administrator | Email message | Mail message about connecting to the system with the transfer of account data and initial password<br>Email message when user resets password | Email | By user request | SMTP |
| EK ASUT | UCP URRAN-Traction | A period of time | The number of repairs, modernizations, and manufactures carried out since the beginning of the calendar year; the number of locomotives received for unscheduled repairs; cases of activation of barrier functions. | Web service | Monthly | JSON // XML |
| ASOUP | UCP URRAN-Traction | A period of time | Date of repair; organizations in which the repairs took place; mileage of locomotives; operational availability coefficient. | SP | Monthly | SESSION table |
| EK ASUTR | UCP URRAN-Traction | A period of time | Injuries occurred in depot; the average number of drivers. | Web service | Monthly | JSON // XML |

- KAS ANT in terms of obtaining information *on technical equipment failures*
- KASAT in terms of obtaining information *on technological violations*
- AS RB in terms of obtaining information *on TSV*
- AS CNSI
- Data warehouse "Locomotive fleet" in terms of obtaining information on: *the volume of work performed by the TRS unit generally; the total mileage of the TRS unit in the context of all types of traffic; linear mileage of head locomotives over the network*
- EC ASUT in terms of obtaining information on: *the number of locomotives received for unscheduled repairs (in the context of depots and regional traction directorates); the number of unit repairs, modernizations and manufactures carried out from the beginning of the calendar year (and by months); cases of activation of barrier functions*
- ASOUP in terms of obtaining information on: *date of repairs of locomotives; organizations in which repairs took place; information about mileage after repair; information about the mileage for the period; acceptable (planned) and actual values of the operational availability coefficient*
- EK ASUTR in terms of obtaining information on: *the number of accidents (resulting in injuries) occurred in the depot (with a breakdown by the severity of injuries); the average number of drivers in the depot; the average number of drivers on the network.*

The interaction of the EK ASUTR and the UCP URRAN-Traction is carried out through the integration mechanisms of ISOP (Integration Service-Oriented Platform). A web service is implemented within ISOP. It submits a request for data to EK ASUTR and returns a response to UCP URRAN-Traction.

Information interaction of the UCP URRAN-Traction system with external components is shown in Fig. 9.8

UCP URRAN-Traction users are:

- Administrators of the Main Computer Center, user administrators and technologists of the Technological Support Centers. The required number of personnel is determined based on the methodological and regulatory documents of the organization operating the system
- Heads and specialists of operational locomotive depots of the Traction Directorate
- Heads and specialists of Regional Traction Directorates
- Heads and specialists of the Central Traction Directorate—JSC "Russian Railways" branch.

The role model of the UCP URRAN-Traction system is given in Table 9.12.

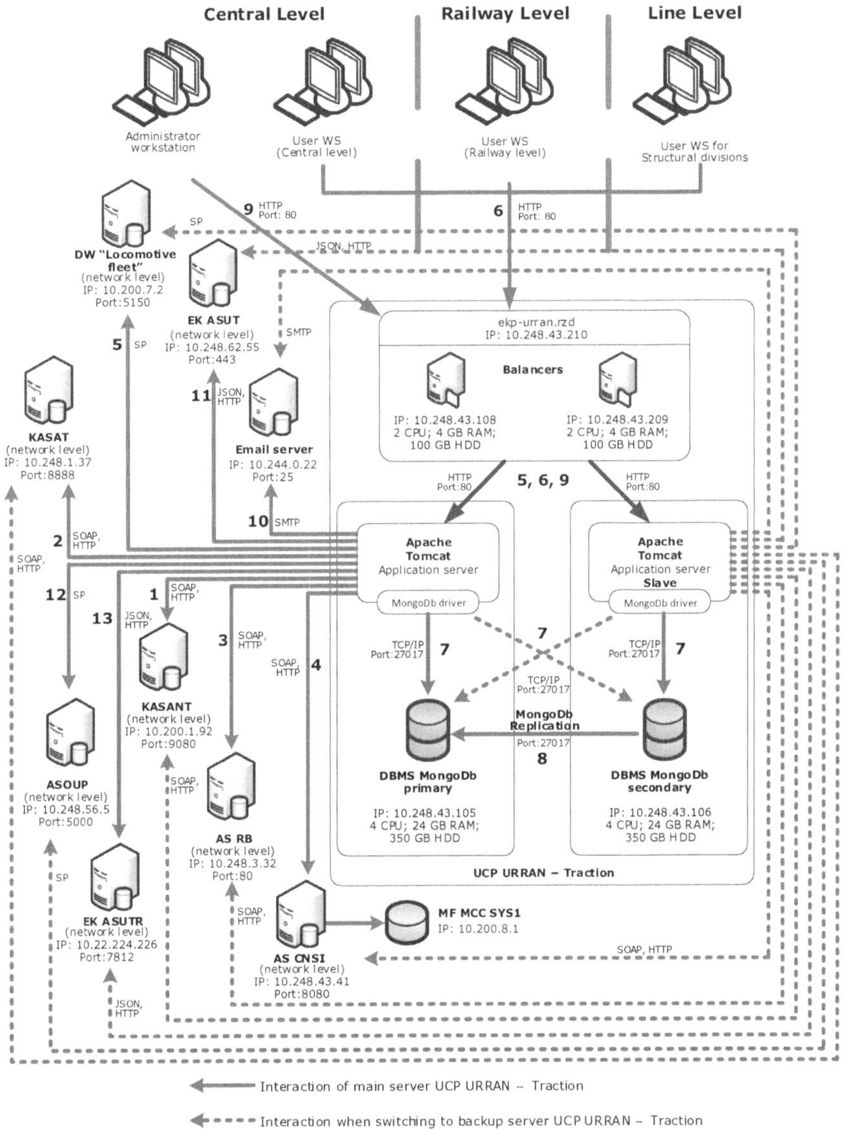


**Fig. 9.8** Scheme of information interaction of UCP URRAN-Traction

**Table 9.12** Role model of the system

| User | Role name | | | | |
|---|---|---|---|---|---|
| | Administrator | Editor of normative reference information | Technology user | Information user | Division manager |
| Staff member of the Main Computer Center and the Technological Support Center | + | – | – | + | – |
| Heads of operational depots | – | – | – | – | + |
| Engineers of the technical sector of the operational depot of Regional Directorate | – | + | + | + | – |
| Heads of Regional Directorates | – | – | – | + | – |
| Engineers of the technical sector of Regional Directorates | – | + | + | + | – |
| Heads of the Traction Directorate | – | – | –– | + | - |
| Engineers of the Technical sector of Traction Directorate | – | + | – | + | – |
| Specialists of Locomotive Engineering and Design Bureau (JSC "Russian Railways" Branch) | – | – | – | + | – |

## 9.8 Prospects for the Development of the UCP URRAN System

The URRAN system is being developed in stages. *At the first* stage, the foundations of the methodology and then, on this basis, the regulatory and methodological support were developed. *At the second stage*, the development and implementation of an automated management system (UCP URRAN) are about to be completed. It is a tool for supporting decision-making on asset management in accordance with the GOST R ISO 55000 "Asset Management" series. *At the third stage*, it is envisaged to introduce (with the help of the UCP URRAN) a technical asset management system in terms of fixed assets of infrastructure complex, locomotive and multiple unit rolling stock complexes at all management levels of the Russian Railways. The efficiency of technical asset management is largely determined by the level of intellectualization of the decision-making support system, including: the ability to adapt to the changing conditions of the technical maintenance of railway transport facilities, the coverage and depth of solving management tasks using artificial intelligence, in particular, using Data Science technology; coverage of risk assessments at all levels of management (beginning with the system, the facility to the process, and ultimately the service).

The listed tasks should mainly be solved *at the fourth stage* of the development of the URRAN system. Let us consider them in more detail.

### 9.8.1 Adaptation Mechanism

It is possible to organize effective management of the technical maintenance of railway transport facilities with the help of the adaptation mechanism. Let us consider a variant of constructing the structure of a system for adaptive management of technical maintenance of objects. This system includes (Fig. 9.9):

- ***An information converter*** (**IC**), which fulfills two groups of tasks: the first group is the connection between the measurable states of the ***X*** system, the non-measurable states ***E,*** and the adaptive action ***S***. By the measurable state is meant the data on the current states of objects (tracks, signalling, communication, power supply, etc.), resource data. Non-measurable states are flows of failures, pre-failures, technological violations, and hazardous events. The *second* group of tasks performed by **IC** is the formation of a vector ***G*** of resource management and control commands ***Y*** to change the current states of a controlled object
- ***Operation (functional) resource*** **L** of the system for the current maintenance of the object

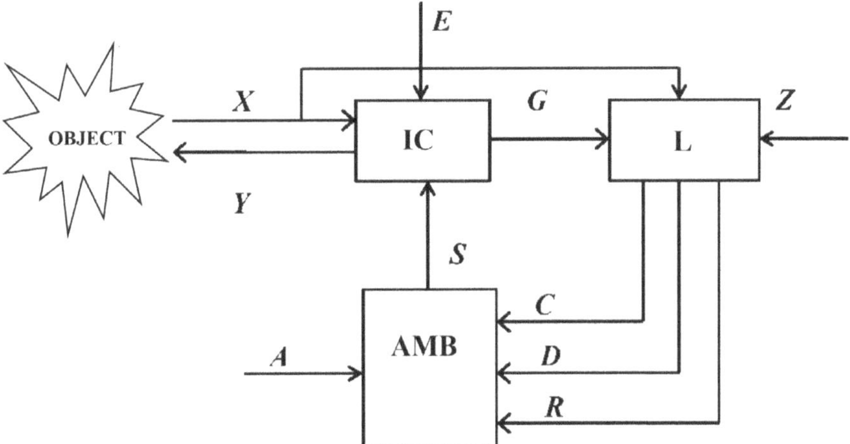


**Fig. 9.9** Structural diagram of the organization of adaptive management of technical maintenance of the railway transport objects

- ***The operator of the adaptive management*** **of AMB**, designed to form an adaptive action in accordance with a certain algorithm ***A***. The adaptation task is to find such a management action ***S*** so that the states of the vector ***G*** satisfy the set goals ***Z***.

The set goals of the technical maintenance for each complex of objects are indicated in item 1.2.1. For example, regarding the track complex, it is assumed that the goal is to reduce the cost of the life cycle of the track infrastructure by redistributing resources, provided that the required level of operational dependability and an acceptable level of train traffic safety are ensured. In this case, the set goals for the track complex are formalized in the following form:

$$Z_1 : \begin{cases} c_j(X,G) \to \text{extr}, j = 1,2,\ \ldots, k_1; \\ d_i(X,G) \geq D_3, i = 1,2,\ \ldots, k_2; \\ r_i(X,G) \leq R_{\text{acceptable}}, i = 1,2,\ \ldots, k_2 \end{cases}$$

where the parameters of the target function of the life cycle cost $c_i$ (for each *j*th management action) are associated with the achievement of the extreme values during adaptation (the minimum life cycle costs of the composite objects of the track complex) provided that a given level of dependability ($D_3$-Dependability) and an acceptable level of safety (which is characterized by the value of the acceptable risk $R_{\text{acceptable}}$) are ensured.

Regarding the complex of signalling and remote control objects, the adopted goal of management of technical maintenance is as follows: increasing the operational dependability of railway signalling and remote control systems while ensuring an acceptable level of train delays and an acceptable life cycle cost based on the redistribution of resources. Therefore, the set goals are formalized in the following form:

$$Z_2 : \begin{cases} d_j(X,G) \to \text{extr}, j = 1,2,\ \ldots, k_1; \\ r_{Ti}(X,G) \geq R_{\text{TD}}, i = 1,2,\ \ldots, k_2; \\ c_i(X,G) \leq C_{\text{Д}}, i = 1,2,\ \ldots, k_2 \end{cases}$$

Now the target function is to achieve the maximum level of dependability of complex objects subject to the constraints regarding the risks of train delays ($R_{\text{TD}}$) and permissible (acceptable) cost of the life cycle of signalling and remote control objects. Goals for other facilities ***Z*** of railway transport are formalized in a similar way.

The minimized (maximized) parameters of the system serve as the goal of adaptation. In addition to the adaptation goal, the adaptive management body (**AMB**) shall also be provided with information on the resource **L** (within which adaptation is possible ($S \subset L$)) and the adaptation algorithm ***A*** which provides the synthesis of the adaptive action ***S*** according to the available information:

$$S = A(X, G, Z, L)$$

Algorithm **A** solves the optimization task. For this purpose, the task of achieving the goals **Z** is reduced to the task of multicriteria optimization. In particular for the track complex

$$c_j(X, G) \rightarrow \text{extr}_{S \subset V}, j = 1, 2, \ldots, k_1$$

where the set $V$ is determined by the condition $S \subset L$ and restrictions $d_i$, $r_i$. When solving the task of multicriteria optimization it is necessary to identify the dependence of the target function $c_j$ on the management ***S*** by direct calculation of the restrictions $d_i$, $r_i$ and the target function [2].

Solving this task of increasing the efficiency of technical maintenance management of all envisaged facilities of the JSC "Russian Railways" should be carried out in a separate module of the UCP URRAN. For this purpose, it is necessary (for each type of object) to regulate the content of each of the vectors ***X, G, Y*** and formalize the optimization algorithms ***A*** using the existing regulatory documents. At this point the maintenance and list of the resource are assessed in the corresponding modules of the UCP URRAN. There are methodologies for normalization of key indicators of dependability and safety. It makes it possible to evaluate the limitations while solving multicriteria optimization task in an automated fashion.

### 9.8.2 Artificial Intelligence

The use of artificial intelligence in the URRAN system is not a fashion statement.

Currently, the following scheme is working: object **operation-event** (for example, failure)–**reaction** (elimination of failure). Based on the currently well-developed artificial intelligence technology Data Sciences, a transition is being made to a progressive technology for managing the technical content of objects according to the scheme: **operation and technical diagnostics–predictive analysis–proactive action**. Predictive analysis is the analysis of current and historical data/events based on mathematical statistics, game theory, etc., to predict data/events in the future.

Proactive action is preliminary work on improving the dependability of those facilities the failures (and especially hazardous failures) of which are predicted. Therefore, there is a critical need for accurate and reliable prediction of non-measurable states of the system (see Fig. 9.9): hazardous failures, pre-failures, failures, and technological violations.

***Data Science*** is a set of concepts and methods of artificial intelligence that allow us to make sense of huge amounts of data and present them in an understandable form.

Data Science is a part of computer science theory that includes methods and algorithms of machine learning. The use of this technology is also effective when supporting management decisions, for example, when determining investments, assessing the activities of structural divisions, assigning repairs, extending the assigned service life of an object, increasing the functional life, etc. It is possible to solve all these tasks using the Data Science technology if there are big arrays of data which are currently processed with the use of Big Data methods (***Big Data*** means unconventional methods of distributed processing of data which volume is more than 1TB, where $1\text{TB} = 2^{10}\ \text{GB} = 2^{20}\ \text{MB}$).

The management of technical maintenance using the UCP URRAN is based on the processing of big arrays of data coming from the complex automated systems: KASANT, KASAT, ASRB, EKASUI, ESMA, AS CNSI, AS ANPSh, EK ASUT, EK ASUMV, EK ASUTR (Unified Corporate Human Resources Management System), EK ASUFR (Unified corporate automated financial and resource management system), ASOUP, DW "Locomotive fleet," regional automated control systems (diagnostic trains, etc.) (see Fig. 7.2). This information is necessary and sufficient to apply artificial intelligence based on Data Science technology. Let us consider it in more detail using the example of solving the task of predicting hazardous events in railway transport.

Repeated passage of trains, high speed of movements on the railway network, environment conditions, and aging cause wear and tear of railways, primarily the railway track. Rail defects may cause derailments or crashes. Such accidents result in damage to the track, power supply systems, as well as wagons and locomotive units with its potential exclusion from the inventory rolling stock.

Derailed units of rolling stock may also intrude into the operational space of the adjacent track, which may cause a collision with an opposing train and, as the consequence, cause catastrophic damage. A significant share of undesired events due to the condition of track is typical not only to Russia's railways. Over the last decade, about a third of all railway incidents in the USA were caused by track-related defects [3].

The analysis of derailments and crashes involving units of freight trains identified that derailments and crashes caused by track malfunctions could occur on a kilometer of track rated, for instance, as "good." In this context, the aggregated estimate of a kilometer of track is not sufficient for predicting its condition, and it is required to take into consideration other parameters: number of widenings, realignments, etc. However, the collection of additional parameters alone will not suffice. Only a part of data on a controlled item is useful in terms of decision-making when managing specific events.

Modern methods of multiple factor data analysis and machine learning technology that allow including over 50 factors into models enable (on the basis of existing values of measured features that characterize the condition of track) making conclusions regarding the need for urgent repairs in order to avoid track failures and train derailments/crashes caused by an unsatisfactory condition of track. Conclusions regarding the efficiency of using Big Data and Data Science technologies can be made on the basis of existing international practical experience, the analysis of which is given below.

Data Science technologies are also extensively used in the railway industry [4–6, etc.]. Machine learning is increasingly popular as means of improving the dependability of railway systems. It also allows minimizing the daily cost of the maintenance [7].

***Machine learning methods*** can be divided into classical algorithms [4] and deep learning methods [7]. The main difference is their presentation level. The classical learning methods include: method of extreme Gradient Boosting XGB, method of Gradient Boosting based on a single leaf decision trees AdaBoost, Support Vectors Method [8], decision trees (e.g., algorithm C4.5 [9]), Random Forest [10], Logit regression [11], and k-Nearest Neighbor algorithm [12], Principal Components Algorithm, etc.

In [3], the principal component analysis in combination with the support vector machine was applied to a dataset of 31 objects collected for the US class I railway network to detect four types of surface defects.

As of late, the academic community has been taking advantage of the deep learning methods to study rail defects. Researchers believe that deep learning may become an element of completely automatic railway monitoring systems [6].

*Deep learning algorithms* based on neural networks are applied as the primary tool to detect structural defects in rails. ***Convolutional neural networks*** (CNNs) are a special case of artificial neural networks. That is due to the widespread use of video cameras that supply the research community with vast quantities of data and enable the application of more complex learning methods.

However, CNN is a "black box" and practically cannot be interpreted. In other words, a researcher of machine learning cannot explain how a CNN model made its predictions or prove their dependability for the end user.

A more detailed overview of the application of various methods of machine learning for detecting track defects can be found in [6].

The diversity of the used models is evidence of the fact that the application of the machine learning technology currently represents a research process that includes the following stages:

- Analysis of the sources of information on the railway track condition
- Preparing data for machine learning
- Definition of machine learning objectives
- Training of models
- Selection of the best model
- Application of the model.

The purpose of the model for the predictive analysis within the track complex consists in predicting the possibility of occurrence of a hazardous failure in the next month within a certain kilometer of the railway track with acceptable levels of accuracy and reliability. The initial information is a set of databases of factors that directly or indirectly affect the dependability of the track complex items. Learning and test samples are formed with the help of databases compiled in automated systems. Using these samples we classify the initial factors.

Sample is one of the key concepts of machine learning. A sample is a finite set of cases (items, instances, events, and test articles) and corresponding data (item characteristics) that form the description of the case. A sample that includes a full set of available data must include the target variable, i.e. an indicator, the prediction of whose value is the goal of machine learning. Additionally, a sample is subdivided into two parts: the learning sample and the test sample.

The tasks of machine learning are normally described in terms of the ways a machine learning system is to process the learning sample. As the case of track upper structure learning sample, we can choose a kilometer of track upper structure (TUS), whose condition is characterized by 77 parameters, including the diagnostic results, operational conditions, and qualitative estimates. The values of such parameters are represented in the form of vector, each element of which is the value of a feature.

Classification is the most common task of machine learning and consists in building models that serve to assign the examined item to one of the several known classes. With respect to that type of tasks the classification algorithm is to answer the question as to which category the item belongs to. In terms of traffic safety (prevention of derailments and other accidents) each kilometer of TUS is divided into two classes: 0, a kilometer with no hazardous TUS failure; 1, a kilometer with a hazardous TUS failure.

From the learning sample we select the best parameters for the classification algorithm. On the test sample we calculate the classification error in order to select the best algorithm.

The task of learning is to re-establish the functional relationship between items and responses, i.e. to construct algorithm that approximates the target relationship in the whole set of possible responses, not only the items of the learning sample.

By the track item is meant not only the physical section of the railway track, but also its condition within one kilometer per month, i.e. a set of data on the track section and its operating and condition parameters. The prediction of the occurrence of a hazardous failure is made for the next month.

The output of the predictive analysis model classifier is not a strict class value, but the confidence in the availability of a "positive" class, which can take values from 0 to 1. For each track item, the classifier calculates the probability that it belongs to a positive class, i.e. it is a track item with a potential hazardous failure. To make a final decision, it is necessary to set a *threshold* of the probability of a hazardous failure. If the probability is less than this threshold, then the track item belongs to the negative class and is assigned the mark "0" (item without hazardous failure). If the probability is greater than the specified threshold value, then the track item is assigned the mark "1" (an object with a potential hazardous failure).

The choice of *the threshold* is based on the following concepts:

1. *The threshold* is 0.5, if the number of "0" and "1" in the area of target feature encoding is identified equally as often and distribution of errors in the data (there is "1" instead of "0" or there is "0" instead of "1") is symmetric

**Table 9.13** Decision threshold of the probability of hazardous failures

| Probability threshold | Goal | Explanation |
|---|---|---|
| threshold 1 | Not to miss a hazardous failure | To predict about 70% of hazardous failures while a sufficiently large number of monitored items will be with false warning of a possible hazardous failure |
| threshold 2 | To make a balanced forecast | Not to miss a hazardous failure and to prevent false prediction of hazardous failure |
| threshold 3 | To prevent false predictions of failure | The share of correctly identified "0" is approximately 95%, but at the same time a fairly large part of hazardous failures will be missed |

2. If it is more important to correctly predict hazardous failures, then *the threshold* should be lowered. If it is more important to prevent the appearance of track items that are mistakenly classified as "1," then *the threshold* should be raised
3. The presence of errors in the data.

Experience has shown that initial data typically does not contain the following data: the columns "track" and/or "kilometer" in respect to the recorded hazardous failures are not filled. This fact does not allow us to correctly mark the analyzed track items. It means, our data contains many track items with hazardous failure, but they were marked "0" instead of "1." The opposite situation is much less common. Therefore, the imbalance (most objects are marked "0") due to the presence of such errors is increased. In this case, *the threshold* should be significantly lowered.

Table 9.13 shows three main values of probability threshold.

The recommended range for choice of the decision *threshold* value is adopted for each polygon of the railway. The principles for choosing *threshold 1, threshold 2, and threshold 3* indicated in Table 9.13 were adjusted taking into account changes in the initial data on the state of the railway track, which affect the accuracy of predicting failures.

Predictive analysis models were created individually for each of the railways. Within each predictive analysis model, the sample is divided into learning and test sample according to the years of observations: the learning sample includes observations of all years except the last one, and the test sample includes observations of the last year.

To overcome the problem of "over learning," as well as to select hyper-parameters of models using training samples, we use a cross-validation procedure with 10 "subsamples." The subsamples are formed in such a way that each of them contains the most different track items by identifiers TMDep, TRACK, KM, and MONTH. For each machine learning model, a set of target hyper-parameters was selected by expertise. This set is to be optimized on a certain grid of parameters.

As the primary output data of the prediction, we use the probabilities of belonging to the classes "0" and "1," which are converted into class marks using a cut-off threshold set by the requirements for accuracy of prediction. The threshold level for

**Table 9.14** Description of samples

| Railway name | Gorkovskaya | Kuibyshevskaya | Northern |
|---|---|---|---|
| Number of track items (before cleansing) | 32,789 (232,658) | 29,030 (89,745) | 100,839 (254,808) |
| Number of features | 193 | 193 | 193 |
| Learning sample | January 2017–December 2018 | January 2019–December 2019 | July 2017–December 2018 |
| Test sample | January 2019–November 2019 | January 2020–February 2020 | January 2019–February 2020 |
| The number of track items in the learning sample | 215,42 | 23,730 | 27,164 |
| The number of track items in the test sample | 11,247 | 5300 | 19,760 |
| The number of hazardous failures in the learning sample | 3874 | 981 | 709 |
| The number of hazardous failures in the test sample | 2771 | 232 | 397 |

each model was set in three ways: to obtain ~ 70% accuracy of identification of class "1" items, to obtain ~ 95% accuracy of identification of class "0" items, and a balanced threshold when the accuracies of identification of items of class "1" and "0" are balanced. Learning and test samples are formed. Table 9.14 shows the main characteristics of the samples for the *threshold 1* target feature in respect to three railways—Gorkovskaya, Kuibyshevskaya, and Severnaya. In regard to the Gorkovskaya Railway, items with following TMDep numbers were excluded from observations: 4, 16, 17, 23, 25, 26, and 27. In regard to the Northern Railway, items with following TMDep numbers were excluded from observations 12, 17, 19, 22, 24, 26, 29, 30, 33, and 36. As for the selected TMDep, the share of records (in the "Failures" database) with key column that are not filled is no more than 15%. As for the Kuibyshevskaya Railway, this share is more than 15% for all TMDep, therefore, all TMDep are included in the sample.

Predictive analysis models for each of the railways were built using the following classification algorithms XGBoost, RForest, SVM, kNN, AdaBoost, and Logit. These algorithms differ in conceptual approaches and mathematical content, which makes it possible to choose the most suitable algorithm for each data sample. For each model, target hyper parameters were selected. These parameters are to be optimized by cross-validation. Table 9.15 shows these parameters for the AdaBoost algorithm.

There are a number of methods to analyze the accuracy of the machine learning algorithm and compare the accuracy of different algorithms of binary classification:

- TP is the number of correctly predicted items marked with "1"

  *Note:* Mark "1" denotes the occurrence of a hazardous failure

- FN is the number of true category "1" items with "0" prediction

**Table 9.15** Optimized hyper-parameters of the AdaBoost model for the polygon of three railways

| Description | Gorkovskaya Railway | Kuibyshev Railway | Northern Railway |
|---|---|---|---|
| Boosting algorithm type | Discrete | Discrete | Discrete |
| Number of boosting iterations | 80 | 82 | 85 |
| Boosting "shrinkage" parameter | 0.12 | 0.11 | 0.12 |
| The parameter of the selection of items characterizing the "randomness" of the choice | 0.49 | 0.5 | 0.52 |
| Flag for using weights in boosting | TRUE | TRUE | TRUE |
| Flag to determine if the weights should be assigned "1" | FALSE | FALSE | FALSE |
| The number of iterations performed at a step of Newton's algorithm to determine the coefficient | 20 | 10 | 10 |
| Lower probability threshold | 0.15 | 0.025 | 0.02 |
| Balanced probability threshold | 0.5 | 0.5 | 0.5 |
| Upper probability threshold | 0.75 | 0.59 | 0.55 |

*Note:* Mark "0" denotes an operable state (no hazardous failure)

- FP is the number of true category "0" items with "1" prediction
- TN *is* the number of correctly predicted category "0" items.

Below are the primary measures of the quality of binary classification models.

1. The General Accuracy of the Algorithm that Defines the Overall Efficiency of the Classifier in Terms of Correct Answers:

$$\mathrm{AC} = \frac{P + \mathrm{TN}}{\mathrm{TP} + \mathrm{FP} + \mathrm{TN} + \mathrm{FN}} \tag{9.1}$$

2. False alarm demonstrating the efficiency of the classifier in terms of predicting deviations from the normal state:

$$\mathrm{FPR} = \frac{\mathrm{FP}}{\mathrm{FP} + \mathrm{TN}} \tag{9.2}$$

3. Accuracy of the algorithm that shows the share of accurately predicted items identified as category "1":

$$\mathrm{PR} = \frac{\mathrm{TP}}{\mathrm{TP} + \mathrm{FP}} \tag{9.3}$$

**Table 9.16** Accuracy indicators for different predictive analysis models on a test sample for all Railways

| Indicators | XGBoost | RForest | SVM | kNN | AdaBoost | Logit |
|---|---|---|---|---|---|---|
| PR | 0.7790 | 0.7772 | 0.6510 | 0.7112 | 0.7511 | 0.7037 |
| FPR | 0.0800 | 0.0802 | 0.0839 | 0.0909 | 0.0799 | 0.0853 |

4. Completeness of the algorithm that shows the share of items that are effectively category "1" and were predicted correctly:

$$\mathrm{RE} = \frac{\mathrm{TP}}{\mathrm{TP} + \mathrm{FN}} \tag{9.4}$$

5. *F*-measure of the algorithm is the harmonic average of accuracy and completeness:

$$F = \frac{2\mathrm{PR} * \mathrm{RE}}{\mathrm{PR} + \mathrm{RE}} \tag{9.5}$$

6. Area under the error curve AUC is a global quality characteristic whose values are between 0 and 1. The value 0.5 corresponds to random guessing, while the value 1 implies correct recognition. AUC indicator is the area under the ROC curve. The ROC curve shows the correlation between the share of false positive classifications (FPR) and share of correct positive classifications (RE). The ROC curve is a sufficiently complex measure of algorithm accuracy. More details about it can be found in [13].

Table 9.16 shows the comparison of the classification methods by the key accuracy indicators *PR* (more is better) and *FPR* (less is better).

The probability *threshold* for all methods was set equal to 0.1 in order to ensure the objectivity of models comparison.

The comparison was made for all Railways on the test sample.

As you can see from Table 9.16, XGBoost, Random Forest, and AdaBoost methods are more accurate than SVM, kNN, and Logit methods. The gradient boosting model based on decision trees—XGBoost—may be considered as the best quality one. This method gives the highest probability of a correct answer with the lowest probability of a false alarm.

It is proposed [13] to use an algorithm, which includes five stages shown in Table 9.17 in order to prepare the data obtained from JSC RZD AMSs.

The results of a numerical experiment of track sections categorization based on failure prediction are presented in [13, 14].

The approbation of the developed models of predictive analysis of hazardous failures of track items was carried out within 8 months of 2020 and showed

**Table 9.17** Stages of data preparation

| Name of stage | Aim | Conditions of stage performance | Relevance criterion of the stage |
|---|---|---|---|
| Data cleansing | Improvement of simulation quality through higher quality of data | Performed always | Performed always |
| Data conversion | Improvement of simulation quality through the capability to compare sequences with different physical units and/or value ranges | Performed if required for discrete sequences | Value variation ranges of various features differ by more than 5 times.<br>Different physical units of features |
| Data sampling | Extension of the scope of applicable models | Performed if required for continuous sequences | Target feature is a continuous value, but it is required to evaluate the probability of being within the range.<br>It is planned to use a method that does not allow using continuous data |
| Text cleansing | Improvement of simulation quality through higher quality of data | Performed if required for continuous sequences | It is planned to use information from the text in the simulation |
| Sampling | Quality verification of the developed models | Performed always | Performed always |

successful results. The prediction accuracy was assessed over the control period. To make forecast for June, we used data for May from the test period, and to make forecast for July, we used data for June.

Table 9.18 shows an example of assessing the quality of predicting hazardous failures with a *threshold 2* (equal chances of getting an error with a mark "0" and a mark "1") for three Railways over the control period. The control period consists of two months of 2020: June and July.

Analysis of Table 9.18 shows that, for example, the number of correctly predicted items of categories marked "1" on the Gorkovskaya Railway is TP = 413, and the number of correctly predicted items of categories marked "0" on the same Railway is TN = 3498. The same table regarding the Gorkovskaya Railway gives the actual data on the falsely predicted marks "0" and "1." As a result of calculations using formulas (9.1)–(9.5), it was found that the efficiency of the classifier in terms of correct answers is AS = 0.77. At the same time the probability of a false alarm is FPR = 0.03. Similar results were obtained for the other two Railways.

Algorithms for classifying item states using various models, an algorithm for assessing the quality of models, a conceptual model algorithm, and other supporting algorithms for implementing the Data Science technology are implemented in a separate UCP URRAN subsystem to be available for all virtual applications.

**Table 9.18** Results of experiments using the XGBoost model for three Railways with a *threshold 2* for the control period

| Description | | | Gorkovskaya Railway | | | Northern Railway | | | Kuibyshevskaya Railway | | |
|---|---|---|---|---|---|---|---|---|---|---|---|
| Matrix for comparing the prediction with the fact of the form | | | | | | | | | | | |
| | '1' | '0' | | '1' | '0' | | '1' | '0' | | '1' | '0' |
| '1' | TP | FN | '1' | 413 | 1031 | '1' | 76 | 1808 | '1' | 168 | 1154 |
| '0' | FP | TN | '0' | 116 | 3498 | '0' | 23 | 6174 | '0' | 57 | 3321 |
| TN—number of correctly predicted '0'<br>FN—number of falsely predicted '0'<br>FP—number of falsely predicted '1'<br>TP—number of correctly predicted '1' | | | '1'—failure<br>'0'—good state | | | '1'—failure<br>'0'—good state | | | '1'—failure<br>'0'—good state | | |
| AC | | | 0.7732 | | | 0.7734 | | | 0.7423 | | |
| TPR | | | 0.2860 | | | 0.0403 | | | 0.1271 | | |
| PR | | | 0.7807 | | | 0.7677 | | | 0.7467 | | |
| FPR | | | 0.0321 | | | 0.0037 | | | 0.0169 | | |
| *F*-measure | | | 0.4187 | | | 0.0767 | | | 0.2172 | | |

### 9.8.3 Hierarchy of Operational Risks

Operational risks include technical, technological, occupational, fire, and environmental risks. These risks should be evaluated for: **1. Elements; 2. Systems; 3. Processes; 4. Services** (Fig. 9.10).

At present, the tasks of assessing the risks of elements and systems have been solved. An assessment method using risk matrices and an integral assessment method was developed. Standards, methods, and recommendations for risk matrix construction were developed. The integral method of risk assessment requires further development and standardization. In general, it can be assumed that the lower levels of the hierarchy of operational risks have been mostly developed and brought up to practical implementation. Unfortunately, a wide range of tasks of assessing operational risks has not been solved. This refers to the tasks of risk aggregation at the level of directorates, their branches and even a number of structural divisions of railway transport. The tasks of assessing complex risks of processes have not been solved. Complex risk means the simultaneous effect of different types of operational risks on the results of the process. This type of tasks is especially relevant when making decisions on managing the technical maintenance of objects (for example, when allocating investments, assessing the activities of structural divisions, assigning repairs, upgrading, increasing the functional life of systems, etc.).

Risk assessment during the process execution can be carried out by various methods: simulation modeling, the field test method, the experimental method by

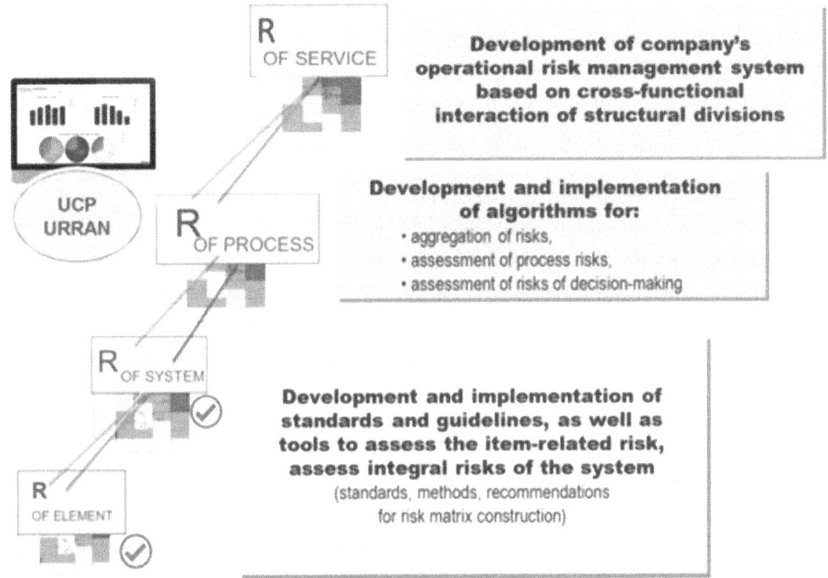


**Fig. 9.10** Hierarchy of operational risks

processing statistical data, the expert method, the experimental and calculation method. All the above methods are reduced to assessing the probability of error in the results of the process and, if the information on the cost of errors is available, they allow using the risk matrix to make a decision on the acceptability (or unacceptability) of the risk due to an error in the results of the process. The use of simulation methods and/or the field test method can be very effective; however, the implementation of these methods is associated with a large amount of preliminary work on the construction and programming of models, which is unacceptable in many cases. The experimental method makes it possible to obtain results that are acceptable in terms of accuracy, provided there is a large amount of statistical data. It is usually impossible to complete in a limited time. The expert method depends on the level of qualifications of experts and gives subjective assessments for the most part. More practical is the experimental and calculation method for assessing the risk of the process. This method can be formed by combining the method for assessing the quality of the technological process [15] and the graph method for assessing the probability of error in the execution of the process algorithm [16]. The combination of these two methods and experimental data on errors in the implementation of compound procedures of technological processes can be the basis for the development of an experimental and calculation method for assessing the risks of industrial processes.

Operational risks in the provision of services are the top of the hierarchy of operational risks. These risks have a great impact on the efficiency of management of the company's technical assets, as well as on the effectiveness of the services provided by the company. Therefore, their importance to the activities of JSC "Russian Railways" is difficult to overestimate. Unfortunately, at present we are not aware of either methodological or practical ideas for managing these risks. One of the possible ways to solve this problem is as follows. We classify possible operational risks of service provision. Then, with assistance of representatives of the company's economic unit, this risk classifier is assessed and supplemented with a column "Amount of damage associated with risks." In parallel, we construct models for calculating and predicting the probability of operational risks. These models can be constructed using the methods of functional dependability theory in combination with the Data Science technology.

To train the models, we should form a database that can be compiled using the output data of several existing AMS, such as KASAT, ASRB, EKASUFR, EK ASUTR, etc. [17]. A research is required to assess the feasibility of this approach. There are other ways to solve the problem of assessing the risk of rendering services by the Company or its branches (subsidiaries). Of course, the careful attention should be paid to solving this problem.

## References


1. Avizienis, A., Laprie, J-C., Randell, B.: Dependability of computer systems. Fundamental concepts, terminology and examples. Technical report. LAAS – CNRS (2000)
2. Podinovsky, V.V., Nogin, V.D.: Pareto-optimal'nye resheniya mnogokriterial'nyh zadach (Pareto-optimal solutions of multicriterion problems). Nauka, Glavnaya redakciya fiziko-matematicheskoj literatury, Moscow (1982)
3. Lasisi, A., Attoh-Okine, N.: Principal components analysis and track quality index: a machine learning approach. Transp. Res. Part C Emerg. Technol. **91**, 230–248 (2018)
4. Thaduri, A., Galar, D., Kumar, U.: Railway assets: a potential domain for big data analytics. Proc. Comput. Sci. **53**, 457–467 (2015)
5. Famurewa, S.M., Zhang, L., Asplund, M.: Maintenance analytics for railway infrastructure decision support. Journal Qual. Maint. Eng. **23**, 310–325 (2017)
6. Nakhaee, M.C., Hiemstra, D., Stoelinga, M., van Noort, M.: The Recent Applications of Machine Learning in Rail Track Maintenance: A Survey. Lecture Notes in Computer Science. 91–105 (2019)
7. Goodfellow, I., Bengio, Y., Courville, A.: Glubokoe obuchenie (Deep Learning). DMK Press, Moscow (2018)
8. Cerrada, M., Zurita, G., Cabrera, D., Sánchez, R.V., Artés, M., Li, C.: Fault diagnosis in spur gears based on genetic algorithm and random forest. Mech. Syst. Signal Process. **70–71**, 87–103 (2016)
9. Santur, Y., Karakose, M., Akin, E.: Random forest based diagnosis approach for rail fault inspection in railways. National Conference on Electrical, Electronics and Biomedical Engineering. 714–719 (2016)
10. Chistyakov, S.P.: Sluchajnye lesa: obzor (Random forests: review.). Trudy Karel'skogo nauchnogo centra RAN. 1, 117–136 (2013)
11. Hosmer, D., Lemeshov, S., Sturdivant, R.X.: Applied logistic regression. Wiley, New York (2013)

12. Hu, C., Liu, X.: Modeling Track Geometry Degradation Using Support Vector Machine Technique. 2016 Joint Rail Conference (2016)
13. Shubinsky, I.B., Zamyshliaev, A.M., Pronevich, O.B., Platonov, E.N., Ignatov, A.N.: Application of machine learning methods for predicting hazardous failures of railway track assets. Dependability Journal. **2**(73), 43–53 (2020)
14. Zamyshliaev, A.M.: Premises of the creation of a digital traffic safety management system. Dependability Journal. **4**(71), 45–52 (2019)
15. Druzhinin, G.V., Sergeeva, I.V.: Kachestvo informacii (The quality of information). Radio i svyaz', Moscow (1990)
16. Shubinsky, I.B., Zamyshlyaev, A.M., Pronevich, O.B.: Graph method for evaluation of process safety in railway facilities. Dependability Journal. **17**(1), 40–45 (2017)
17. Zamyshlyaev, A.M.: Prikladnye informacionnye sistemy upravleniya nadezhnost'yu, bezopasnost'yu, riskami i resursami na zheleznodorozhnom transporte (Applied information systems for management of dependability, safety, risks and resources in railway transport). LLC "Journal Dependability", Moscow (2013)

# Chapter 10
# Conclusion

Technical assets bring profit to a company only if they operate dependably (uninterruptedly) and efficiently. The efficiency of an asset is directly correlated to its productivity, level of functionality, and especially safety.

Acceptable levels of dependability and, largely, safety are achieved on the basis of a rational system of technical maintenance of an object (technical asset of a company). The required levels of safety and dependability of facilities, as well as sufficient levels of their productivity and functionality, while ensuring an acceptable cost of the facility's life cycle, should be achieved by a balanced risk-based management of resources. The solution to this problem is especially relevant for such a large backbone transport company as JSC "Russian Railways," which has an extremely large number of technical assets of infrastructure and rolling stock.

The management of technical assets of JSC "Russian Railways" is carried out on the basis of the URRAN system (Resource and risk management by analysis of and ensuring dependability and safety). This system is composed of three interrelated components:

- Risk-based methodology for adaptive management of the technical maintenance of railway transport facilities, the activities of structural divisions, dependability and safety of the transportation process;
- Regulatory and methodological framework of the system
- Informatization of the processes of data collection and processing, management of technical assets; automation of all regulatory documents developed within the URRAN system.

Each complex of the facilities of JSC "Russian Railways" has specific features, which are due to the role of this complex within the transportation process, the conditions of its performing, and the established relationships with other complexes. Therefore, the goals of introduction of the URRAN system are specific for each complex of facilities. For example, regarding the track complex, it is assumed that the goal is to reduce the cost of the life cycle of the track infrastructure by



I. B. Shubinsky, A. M. Zamyshlaev, *Technical Asset Management for Railway Transport*, International Series in Operations Research & Management Science 322,
https://doi.org/10.1007/978-3-030-90029-8_10

redistributing resources, provided that the required level of operational dependability and an acceptable level of train traffic safety are ensured. As for the complex of signalling and remote control facilities, the adopted goal of managing technical maintenance is significantly different. It is as follows: increasing the operational dependability of railway signalling and remote control systems while ensuring an acceptable level of train delays and an acceptable life cycle cost based on the redistribution of resources.

To accomplish the envisaged goals, the URRAN system solves the following tasks:

- It allows assessing and predicting the indicators of dependability and safety of infrastructure and rolling stock facilities (including those with complex excessive structures) in real time;
- Management of technical, technological, fire, occupational, and environment risks;
- Assessment of wear, residual lifetime, and the limit state of railway transport facilities;
- Prediction of the state of infrastructure facilities. Prediction of hazardous track failures;
- Assessment of the life cycle costs of railway transport facilities;
- Assessment of the activities of divisions of JSC "Russian Railways" taking into account results of their activities on ensuring the dependability and safety of the facilities operated;
- Management of resources used for technical maintenance;
- Providing decision-making support on the basis of a unified corporate platform URRAN.

To solve these tasks, the URRAN methodology includes a set of applied analysis methods, as well as a number of ingenious methods that allow, for the first time, solving a number of non-trivial tasks associated with the management of technical assets. They include: semi-Markov graph methods for calculating and predicting the functional safety of transport facilities, a matrix method for assessing the risks of objects, a matrix method of the integrated assessment of system risks, a method of control and assessment charts, a method for supporting decision-making on extending the service life of an object, and risk-based methods of score assessment of the division activities. Regulatory and methodological framework was developed on the basis of the URRAN methodology and includes about 150 documents (GOSTs, GOST R, STO RZD, classifiers, etc.) regulating various aspects of asset management and activities of branches.

Currently, the following regulatory and methodological documents intended for the management of technical assets of infrastructure facilities complex and rolling stock complex were automated:

- Documents in the field of dependability and safety, risk management;
- Documents regarding the assessment of the structural division activities;

- Documents regarding the assessment of physical wear, residual life, and functional life;
- Documents regarding analysis of pre-failure states of objects;
- Documents in the field of economics and planning of the repair;
- Documents in the field of traffic safety;
- Documents in the field of labor protection;
- Documents regarding the assessment of occupational risk;
- Documents in the field of fire safety.

These tasks were solved within the information and management system Unified Corporate Platform (UCP) URRAN.

The UCP URRAN system contains five functionally complete *virtual technical subsystems*: UCP URRAN-Track (automation within the complex of track upper structure and structures), UCP URRAN-Signalling (automation within the signalling and remote control complex), UCP URRAN-Energy (automation of railway power supply and electrification facilities), UCP URRAN-Communication (automation within the railway telecommunication complex), UCP URRAN-Traction (automation of complexes of locomotive and multi-unit rolling stock). Subsystems interact via programming interfaces.

The effectiveness of technical asset management is largely determined by the level of intellectualization of the decision-making support system, including: the ability to adapt to the changing conditions of the technical maintenance of railway transport facilities, the scope and depth of solving management problems using artificial intelligence (in particular, Data Science technology), and coverage of risk assessments at all levels of management (from system, facility to process, and ultimately service). The listed tasks should mainly be solved at the next stage of development of the UCP URRAN. The use of artificial intelligence in the URRAN system is not a fashion statement. The following procedure is being currently used: *operation* of an object–*event* (for example, failure)–*action* (elimination of failure), but at the same time there is a tendency to migration toward the advanced technology of managing technical maintenance of the object on the basis of well-developed technology of artificial intelligence Data Sciences according to the following procedure: *operation and technical diagnostics–predictive analysis–proactive action.* Predictive analysis is the analysis of current and historical data/events (based on mathematical statistics, game theory, etc.) to predict data/events in the future. Proactive action is preliminary work on improving the dependability of those facilities failures (and especially hazardous failures) of which are predicted. Therefore, there is a critical need for accurate and reliable prediction of non-measurable states of the system (see Fig. 9.9): hazardous failures, pre-failures, failures, and technological violations. A large volume of statistical data on technological violations in transport makes it possible to form representative learning samples and, on this basis, predict possible technological violations committed by specialists of infrastructure complex and rolling stock complex.

The manufacturer's authorised representative in the EU is Springer Nature Customer Service Centre GmbH, Europaplatz 3, 69115 Heidelberg, Germany. If you have any concerns regarding our products, please contact ProductSafety@springernature.com

Printed and bound by CPI Group (UK) Ltd, Croydon, CR0 4YY

29/04/2026

02099522-0002